Multilateral Environmental Agreements and Compliance

The adoption of administrative procedures in global governance has the potential to foster proper consideration of marginalized actors' interests, yet risks entrenching the dominance of the well-resourced and powerful. Accordingly, this book proposes a new framework for evaluating the extent to which administrative procedures in the compliance systems of multilateral environmental agreements constrain power and promote regard for the interests of affected states, which are frequently developing and transition countries. This framework is applied to the compliance systems under the Montreal Protocol, the Kyoto Protocol and the Convention on International Trade in Endangered Species of Wild Fauna and Flora (CITES), which address critical global environmental issues of ozone-layer depletion, climate change and trade in endangered species, respectively. The analysis shows that, under certain conditions, administrative procedures limit the influence of states' asymmetric power on compliance deliberations. Furthermore, systematic adoption of these procedures increases the opportunities for affected states' interests to be voiced and considered in compliance decision-making processes.

Anna Huggins is a Senior Lecturer in the Faculty of Law, Queensland University of Technology (QUT).

Routledge Research in International Environmental Law

Titles in this series include:

Ecological Integrity and Global Governance
Science, Ethics and the Law
Laura Westra

Environmental Law and the Ecosystem Approach
Maintaining Ecological Integrity through Consistency in Law
Froukje Maria Platjouw

Stratospheric Ozone Damage and Legal Liability
US Public Policy and Tort Litigation to Protect the Ozone Layer
Lisa Elges

Ecological Restoration in International Environmental Law
Anastasia Telesetsky, An Cliquet and Afshin Akhtar-Khavari

Sustainable Development Principles in the Decisions of International Courts and Tribunals
1992–2012
Marie-Claire Cordonier Segger and H.E. Judge C.G. Weeramantry

Culture and International Economic Law
Valentia Vadi and Bruno de Witte

International Natural Resources Law, Investment and Sustainability
Shawkat Alam, Jahid Hossain Bhuiyan and Jona Razzaque

Protecting Traditional Knowledge
The WIPO Intergovernmental Committee on Intellectual Property and Genetic Resources, Traditional Knowledge and Folklore
Edited by Daniel F. Robinson, Ahmed Abdel-Latif and Pedro Roffe

Environmental Governance and Common Pool Resources
Michael G. Faure, Peter Mascini and Jing Liu

Multilateral Environmental Agreements and Compliance
The Benefits of Administrative Procedures
Anna Huggins

Multilateral Environmental Agreements and Compliance

The Benefits of Administrative Procedures

Anna Huggins

Routledge
Taylor & Francis Group
LONDON AND NEW YORK

First published 2018 by Routledge

2 Park Square, Milton Park, Abingdon, Oxfordshire OX14 4RN
52 Vanderbilt Avenue, New York, NY 10017

Routledge is an imprint of the Taylor & Francis Group, an informa business

First issued in paperback 2019

British Library Cataloguing-in-Publication Data
A catalogue record for this book is available from the British Library

Library of Congress Cataloging-in-Publication Data
Names: Huggins, Anna, author.
Title: Multilateral environmental agreements and compliance : the benefits of administrative procedures / Anna Huggins.
Description: Abingdon, Oxon [UK] ; New York : Routledge, 2018. | Series: Routledge research in international environmental law | Includes bibliographical references and index.
Identifiers: LCCN 2017027288 | ISBN 9781138288638 (hbk)
Subjects: LCSH: Environmental law, International. | Environmental protection—International cooperation. | Administrative procedure. | Environmental policy—International cooperation. | Vienna Convention for the Protection of the Ozone Layer (1985 March 22). Protocols, etc. (1987 September 15) | United Nations Framework Convention on Climate Change (1992 May 9). Protocols, etc. (1997 December 11) | Convention on International Trade in Endangered Species of Wild Fauna and Flora (1973 March 3) | Climatic changes—Law and legislation. | International economic relations—Environmental aspects.
Classification: LCC K3585 .H84 2018 | DDC 344.04/6—dc23
LC record available at https://lccn.loc.gov/2017027288

ISBN: 978-1-138-28863-8 (hbk)
ISBN: 978-0-367-88643-1 (pbk)

Typeset in Galliard
by Keystroke, Neville Lodge, Tettenhall, Wolverhampton

Table of contents

List of figures

List of abbreviations

CBDR:	Common but differentiated responsibilities
CEITs:	Countries with economies in transition
CFCs:	Chlorofluorocarbons
CITES:	Convention on International Trade in Endangered Species of Wild Fauna and Flora
CMP:	Conference of the Parties serving as the Meeting of the Parties
COP:	Conference of the Parties
ERTs:	Expert review teams
GAL:	Global administrative law
GHG:	Greenhouse gas
HCFCs:	Hydrochlorofluorocarbons
HFCs:	Hydrofluorocarbons
IAR:	International assessment and review
ICA:	International consultation and analysis
IEL:	International environmental law
IUCN:	International Union for the Conservation of Nature
MEA:	Multilateral environmental agreement
MOP:	Meeting of the Parties
NDC:	Nationally determined contribution
NGO:	Non-governmental organization
SBI:	Subsidiary Body for Implementation
TRAFFIC:	Trade Records Analysis of Fauna and Flora in Commerce
TWAIL:	Third world approaches to international law
UAE:	United Arab Emirates
UN:	United Nations
UNDP:	United Nations Development Program
UNEP:	United Nations Environment Programme
UNFCCC:	United Nations Framework Convention on Climate Change
WWF:	World Wide Fund for Nature International

Acknowledgements

This book builds upon my doctoral dissertation, and primary thanks are due to my supervisors at the University of New South Wales (UNSW) Law School. Professor Rosemary Rayfuse was an expert guide in international law and international environmental law, and has greatly inspired my research interest in these areas. This book adapts and applies insights from administrative law and regulatory theory to the international law context, and I could not have asked for better intellectual guidance and inspiration in this regard than I received from Professor Bronwen Morgan. I am indebted to both Professors Rayfuse and Morgan for their kindness, support, generosity and mentorship both during and after my PhD candidacy. My PhD research was also enriched by the excellent guidance provided by the panelists for my progress reviews at UNSW Law, Professors Martin Krygier, Mark Aronson and Colin Picker. I am also grateful to Professors Ellen Hey and Alexander Zahar, both of whom provided valuable perspectives and feedback on the case study on the Kyoto compliance system.

My PhD examiners, Professors Liz Fisher at the University of Oxford and Jaye Ellis at McGill University, were exceptionally generous in providing detailed and insightful comments that helped me to conceptualize this book. The three anonymous reviewers of my book proposal also provided positive, thoughtful and highly constructive feedback, which significantly improved this book. My thinking on the subject matter of this work was also enriched by discussions with Professor Benedict Kingsbury and participants at the Annual Junior Faculty Forum for International Law at New York University (NYU) Law School in June 2016. I would also like to thank Professor Richard Johnstone and Dr Kylie Pappalardo for their collegiality and generosity in providing feedback on parts of my draft manuscript. The editorial team at Routledge has been highly professional and supportive, and it has been a pleasure to work with them throughout the publication process.

I wish to acknowledge that some of the chapters of this book are based on other publications of mine. Parts of Chapter 5 build upon work that previously appeared as 'The Desirability of Depoliticization: Compliance in the International Climate Regime' (2015) 4(1) *Transnational Environmental Law* 101. The penultimate section of Chapter 7 analyzing the compliance framework under the Paris Agreement is informed by my co-authored work with Dr Saiful Karim in 'Shifting Traction: Differential Treatment and Substantive and Procedural Regard in the International Climate Change Regime' (2016) 5 *Transnational Environmental Law 427*.

My family has always strongly supported my academic pursuits, including the writing of this book, for which I am very grateful. In particular, I wish to thank my mother for nurturing my love of writing, and for her thoughtful editorial suggestions on the draft manuscript. My greatest expression of gratitude is reserved for my husband David, to whom I dedicate this book.

Unless otherwise indicated, the law as stated in the text is current as at 1 May 2017.

1 Administrative procedures in MEA compliance systems

1 Introduction

Climate change, ozone-layer depletion and trade in endangered species of fauna and flora are three of the most pressing global environmental issues of our time. The question of how the conflicting interests of states can be acknowledged and considered in international decision-making processes needs to be addressed if effective collective action is to be taken. This book analyzes the role of procedures in enhancing compliance processes in three major global multilateral environmental agreements (MEAs). A key aspect of this inquiry is developing and applying a framework for evaluating the extent to which the adoption of administrative procedures[1] constrains power politics and enhances regard for the interests of affected states in compliance proceedings. This focus on the potential benefits of proceduralization[2] provides a new angle for analyzing compliance in international environmental law (IEL). I argue that administrative procedures that help to level the playing field for less powerful states by giving them greater voice in compliance deliberations, increase the responsiveness of decision making to states' diverse capacities and circumstances, and justify the imposition of consequences for non-compliance, are invaluable elements in the design of compliance frameworks in IEL. Such procedural safeguards are important to provide a counterbalance to asymmetric power relations between states. Given the high stakes of failed international co-operation to address issues of global environmental concern, it is

1 'Administrative procedures' and 'administrative mechanisms', which are used interchangeably in this book, refer to principles and practices of accountability, transparency, participation, reason giving and review from global administrative law: see, eg, B Kingsbury, N Krisch and R B Stewart, 'The Emergence of Global Administrative Law' (2005) 68 *Law and Contemporary Problems* 15, 17. See further Chapter 2, section 3.

2 In this book, 'proceduralization' refers to the trend toward adopting procedural tools of an administrative character in global environmental governance. My definition is similar to, but goes beyond, that of Howarth, who describes proceduralization as 'the imposition of constraints upon those entrusted with environmental decision making': W Howarth, 'Aspirations and Realities under the Water Framework Directive: Proceduralisation, Participation and Practicalities' (2009) 21(3) *Journal of Environmental Law* 391, 398.

important to better understand the optimal design and operation of compliance procedures.

MEA compliance systems oversee states' implementation of their commitments under IEL. These systems encompass states' provision of performance-review information, multilateral non-compliance procedures and non-compliance response measures.[3] Compliance systems in MEAs developed in response to the limitations of adversarial channels of bilateral dispute settlement in international law for addressing multilateral environmental concerns.[4] Accordingly, they reflect a preference for administrative and political, rather than legalistic, compliance solutions.[5] This book focuses on three global MEAs – the Montreal Protocol on Substances that Deplete the Ozone Layer[6] (the Montreal Protocol), the Kyoto Protocol to the United Nations Framework Convention on Climate Change[7] (the Kyoto Protocol), and the Convention on International Trade in Endangered Species of Wild Fauna and Flora[8] (CITES).

While much of the compliance scholarship in international law explores how to enhance states' compliance with their legal obligations, this book provides a new perspective by focusing on the procedural rigor of the decision-making process that may lead to a finding of non-compliance. Two influential schools of thought on promoting compliance with international legal rules are the managerial model[9] and the enforcement model.[10] The managerial model proposed by Chayes and Handler

3 G Rose, 'Compliance with Multilateral Environmental Agreements' in L Paddock et al (eds), *Compliance and Enforcement in Environmental Law: Toward More Effective Implementation* (Edward Elgar, 2011) 3, 5–10. Rose includes 'dispute settlement procedures' as a fourth compliance mechanism; however, as such procedures are judicial rather than administrative in character, and have not been used in the case study multilateral environmental agreements (MEAs) in this book, they are not a focus here. For a similar definition, see S Borras, 'Comparative Analysis of Selected Compliance Procedures under Multilateral Environmental Agreements' in S Maljean-Dubois and L Rajamani (eds), *Implementation of International Environmental Law* (Martinus Nijhoff Publishers, 2011) 319, 366.

4 See, eg, G Ulfstein, 'Dispute Resolution, Compliance Control and Enforcement in International Environmental Law' in G Ulfstein, T Marauhn and A Zimmermann (eds), *Making Treaties Work: Human Rights, Environment and Arms Control* (Cambridge University Press, 2007) 115, 119, 125, 128–9.

5 D Bodansky, *The Art and Craft of International Environmental Law* (Harvard University Press, 2009) 248.

6 *Montreal Protocol on Substances that Deplete the Ozone Layer 1987*, opened for signature 16 September 1987, 1989 UNTS 28 (entered into force 1 January 1989) ('Montreal Protocol').

7 *Kyoto Protocol to the United Nations Framework Convention on Climate Change*, opened for signature 11 December 1997, 37 ILM 22 (entered into force 16 February 2005) ('Kyoto Protocol').

8 *Convention on International Trade in Endangered Species of Wild Fauna and Flora 1973*, opened for signature 3 March 1973, 993 UNTS 243 (entered into force 1 July 1975) ('CITES').

9 See A Chayes and A Handler Chayes, *The New Sovereignty: Compliance with International Regulatory Agreements* (Harvard University Press, 1998).

10 G W Downs, D M Rocke and P N Barsoom, 'Is the Good News about Compliance Good News about Cooperation?' (1996) 50(3) *International Organization* 379.

Chayes assumes that states enter into international commitments with good intentions, and that non-compliance stems from norm ambiguities or capacity limitations, rather than willful disregard.[11] Accordingly, they propose a range of positive measures – specifically, transparency, dispute settlement and capacity building – to 'manage' the causes of non-compliance, which they view as preferable to heavy-handed enforcement responses.[12] The 'enforcement model' approach to compliance is typically associated with the work of Downs, Rocke and Barsoom,[13] although these authors do not describe their work in this way. A focus of this approach is on the use of sanctions, which encompass a range of measures that create costs or remove benefits, and are desirable in situations in which treaties require 'deep co-operation' and there are strong incentives for non-compliance.[14] These two theories sit alongside other explanations of the 'compliance pull' of international law,[15] and why states comply with international legal obligations.[16] Unlike these approaches, the present study does not seek to propose principles and practices for promoting states' compliance with their international environmental commitments. Rather, the focus is on understanding the implications of procedural design in MEA compliance frameworks.

Procedural design warrants further attention due to the risk that, without adequate procedural safeguards, compliance decisions will be strongly shaped by differences in power between states. In the selected MEA compliance systems, states create and enforce compliance rules, and are the subjects of compliance decisions. In such state-centric settings, asymmetric power relations can strongly influence decision-making processes, which risks disadvantaging less powerful developing states.[17] MEA compliance systems operate against a backdrop of

11 Chayes and Handler Chayes, *The New Sovereignty*, above n 9, 10–15.

12 *Ibid* 22–25.

13 Downs, Rocke and Barsoom, 'Is the Good News about Compliance Good News about Cooperation?', above n 10.

14 That is, the treaty requires states to depart significantly from what they would have done in the absence of the treaty: *ibid* 387–392.

15 See, eg, T Franck, *Fairness in International Law and Institutions* (Oxford University Press, 1995) 26; J Brunnée and S J Toope, *Legitimacy and Legality in International Law: An Interactional Account* (Cambridge University Press, 2010) 92.

16 On consent-based theory, see, eg, J K Setear, 'An Iterative Perspective on Treaties: A Synthesis of International Relations Theory and International Law' (1996) 139 *Harvard International Law Journal* 147. On transnational legal process, see, eg, H H Koh, 'Transnational Legal Process' (1996) 75 *Nebraska Law Review* 181. Theories of compliance informed by international economic law include: A T Guzman, 'A Compliance-Based Theory of International Law' (2002) 90 *California Law Review* 1823, 1874; A T Guzman, *How International Law Works: A Rational Choice Theory* (Oxford University Press, 2008).

17 A Tanzi and C Pitea, 'Non-Compliance Mechanisms: Lessons Learned and the Way Forward' in T Treves et al (eds), *Non-Compliance Procedures and Mechanisms and the Effectiveness of International Environmental Agreements* (TMC Asser Press, 2009) 569, 573. Although it is acknowledged that the distinction between developed and developing countries is in some cases becoming less clear-cut, the general proposition that states that do not have developed economies, and particularly small island developing states and least developed countries, are

North–South power imbalances, which reflect the disempowering impacts and ongoing legacies of the colonial encounter, as well as the historical and increasing material and political inequalities between states.[18] Moreover, IEL decision-making contexts traditionally rely on negotiation, diplomacy and co-operation to promote the achievement of community interests in addressing global environmental problems.[19] In the absence of robust due-process guarantees, such negotiation-based practices are likely to favor developed states with greater bargaining power. This can manifest in, for example, developing country Parties being the targets of compliance proceedings and trade sanctions, while powerful developed countries that commit similar infractions are not similarly targeted.

This book explores if and how adopting administrative procedures (1) ameliorates the risk of compliance decisions being strongly influenced by power politics and (2) promotes regard for affected states' interests. These are not the only normative aims that are associated with proceduralization.[20] They are, however, two aspirations that are highly relevant to the theme of redressing power imbalances between states in international decision-making processes. Stewart analyzes the role of administrative procedures in disciplining political influences and providing opportunities for previously disregarded interests to be voiced and heard, with a focus on marginalized individuals and groups in global regulatory governance.[21] I argue that

likely to be politically and economically disadvantaged to varying degrees vis-à-vis their developed country counterparts, remains pertinent. Such disadvantage is likely to affect states' compliance with MEAs, as well as their engagement with MEA compliance systems, and will be most pronounced for small island developing states and least developed countries.

18 A Anghie, *Imperialism, Sovereignty and the Making of International Law* (Cambridge University Press, 2005); L Rajamani, *Differential Treatment in International Environmental Law* (Oxford University Press, 2006) 2–3. See generally S Alam, S Atapattu, C G Gonzalez and J Razzaque (eds), *International Environmental Law and the Global South* (Cambridge University Press, 2015).

19 See, eg, S Bankobeza, E Maruma Mrema and K Ramakrishna, 'International Environmental Diplomacy and Negotiations' in T Kuokkanen, E Couzens, T Honkonen and M Lewis (eds), *International Environmental Law-making and Diplomacy: Insights and Overviews* (Routledge, 2016) 83; A Cardesa-Salzmann, 'Constitutionalising Secondary Rules in Global Environmental Regimes: Non-Compliance Procedures and the Enforcement of Multilateral Environmental Agreements' (2011) 24(1) *Journal of Environmental Law* 103, 111–113; Chayes and Handler Chayes, above n 9, 201–228.

20 For example, Kingsbury, Krisch and Stewart identify three normative foundations for global administrative law (GAL): 'international administrative accountability', protection of rights of private or state actors and 'promotion of democracy': Kingsbury, Krisch and Stewart, 'The Emergence of Global Administrative Law', above n 1, 43. Others claim that promoting the rule of law is an aspiration of GAL: see, eg, R B Stewart, 'Remedying Disregard in Global Regulatory Governance: Accountability, Participation, and Responsiveness' (2014) 108 *American Journal of International Law* 211, 220; E Benvenisti, *The Law of Global Governance* (Hague Academy of International Law, 2014) 89–99.

21 R B Stewart, 'Part I Courts, Institutions, and Access to Justice: Legitimacy and Accountability in Global Regulatory Governance: Global Administrative Law and Developing Countries' (2009) 1 *Jindal Global Law Review* 41, 43; Stewart, 'Remedying Disregard in Global Regulatory Governance', above n 20, 244–268. These questions are vitally important as the

a similar logic ought to apply to *states* affected by decisions in MEA compliance systems. If such aims are realized, administrative procedures can enhance the procedural rigor of decision-making processes and facilitate opportunities to be heard for developing and transition countries, which are typically the subjects of compliance proceedings in IEL.[22] This is important as decision-making processes that reflect North–South biases and lack procedural fairness risk alienating less powerful states from global environmental governance.[23] Such alienation needs to be avoided as the political and environmental success of MEAs and their compliance regimes requires commitment from all state Parties.

However, adopting administrative procedures is not necessarily a panacea to states' unequal bargaining power as these procedures may be co-opted by the powerful and well-resourced for their own advantage.[24] It is thus unclear whether adopting administrative procedures does in fact institutionalize regard for affected interests in compliance decisions, or merely entrenches and perpetuates existing power imbalances. In the light of these uncertainties, a central focus of this book is developing and applying a new framework for evaluating the potential benefits of proceduralization in compliance proceedings in IEL.

2 Evaluating the benefits of proceduralization

The uncertain effect of adopting administrative procedures raises the question – how do we know if proceduralization constrains power and enhances consideration of affected interests in MEA compliance decisions? IEL has an increasingly administrative character, yet there is a paucity of systematic analysis of the extent and significance of this trend. For example, Tanzi and Pitea claim that MEA compliance procedures are analogous to administrative procedures.[25] Scott comments that

issues of 'how to tame power, level the institutional playing field, and ensure that all affected interests are adequately represented or at least taken into account' are key issues confronting global governance: Benvenisti, *The Law of Global Governance*, above n 20, 78.

22 See, eg, Chapters 4 and 6. As Brunnée notes, it is likely that non-complying states will have genuine capacity limitations: J Brunnée, 'Enforcement Mechanisms in International Law and International Environmental Law' in U Beyerlin, P-T Stoll and W Rudiger (eds), *Ensuring Compliance with Multilateral Environmental Agreements: A Dialogue Between Practitioners and Academia* (Brill Martinus Nijhoff Publishers, 2006) 1, 19. See also Chayes and Handler Chayes, *The New Sovereignty*, above n 9, 14.

23 E Hey, 'International Institutions' in D Bodansky, J Brunnée and E Hey (eds), *The Oxford Handbook of International Environmental Law* (Oxford University Press, 2007) 749, 768–769.

24 Kingsbury, Krisch and Stewart, 'The Emergence of Global Administrative Law', above n 1, 51–52; Stewart, 'Part I Courts, Institutions, and Access to Justice', above n 21, 43–44; B S Chimni, 'Co-option and Resistance: Two Faces of Global Administrative Law' (2005) 37 *New York University Journal of International Law and Politics* 799, 826.

25 Tanzi and Pitea, 'Non-Compliance Mechanisms', above n 17, 580. In a related vein, Marauhn outlines the emergence of a procedural law of compliance in IEL: T Marauhn, 'Towards a Procedural Law of Compliance Control in International Environmental Relations' (1996) 56(3) *Zeitschrift für Ausländisches Recht und Völkerrecht* 696, 722–730.

'surprisingly, leading international administrative lawyers have paid little attention to the institutions established by international environmental agreements in the development of their sub-discipline to date', notwithstanding that the aptness of their observations to MEA compliance processes is 'undeniable'.[26] While these and other scholars have noted the administrative character of procedures in global environmental governance,[27] the degree to which specific administrative mechanisms have gained traction in MEA compliance systems,[28] and the implications of these developments for affected states, remain under-explored.[29] This book addresses this important research gap.

The state-centric nature of MEA compliance decisions raises a host of different issues to those evident in domestic administrative law systems. Administrative law norms of due process and procedural fairness are accepted in all Western legal systems, and recognized in many others.[30] Although the precise configuration of principles and practices varies in each legal system, common attributes generally include impartiality, the provision of a hearing and a concomitant opportunity to be heard, and reasoned decisions.[31] Due-process guarantees in domestic legal systems seek to redress the power imbalance between individual citizens and the state.[32] Yet there are of course significant differences between domestic and

26 K N Scott, 'Non-Compliance Procedures and Dispute Resolution Mechanisms under International Environmental Agreements' in D French, M Saul and N D White (eds), *International Law and Dispute Settlement: New Problems and Techniques* (Hart Publishing, 2010) 225, 230.

27 See, eg, E Hey, 'The Climate Change Regime: An Enviro-Economic Problem and International Administrative Law in the Making' (2001) 1 *International Environmental Agreements: Politics, Law and Economics* 75; B Kingsbury, 'Global Environmental Governance as Administration: Implications for International Law' in D Bodansky, J Brunnée and E Hey (eds), *The Oxford Handbook of International Environmental Law* (Oxford University Press, 2007) 63.

28 For notable exceptions, see Scott, 'Non-Compliance Procedures and Dispute Resolution Mechanisms', above n 26; and D H Rached, *The International Law of Climate Change and Accountability* (PhD thesis, the University of Edinburgh, 2013) Ch 5.

29 The present focus on affected states, which are typically developing and transition countries, helps to address a lacuna in the GAL literature identified by Urueña Hernández regarding 'the effects of global regulatory governance on the states of the Global South as international legal persons, and their ability to operate in global governance institutions': R F Urueña Hernández, 'Global Administrative Law and the Global South' in S Cassese (ed), *Research Handbook on Global Administrative Law* (Edward Elgar, 2016) 392, 394.

30 C Harlow, 'Global Administrative Law: The Quest for Principles and Values' (2006) 17(1) *The European Journal of International Law* 187, 204. For a comparative study, see, eg, D J Galligan, *Due Process and Fair Procedures: A Study of Administrative Procedures* (Clarendon Press, 1996).

31 M D Bayles, *Procedural Justice: Allocating to Individuals* (Kluwer International, 1990) Chs 2–4.

32 Some of the international law scholarship on due process also focuses on individuals. For example, Hovell insightfully analyzes the values associated with due process for individuals affected by Security Council sanctions decision-making: see, eg, D Hovell, *The Power of Process: The Value of Due Process in Security Council Sanctions Decision-Making* (Oxford University Press, 2016); D Hovell, 'Due Process in the United Nations' (2016) 110(1) *American Journal of International Law* 1.

international legal frameworks and institutions, including the absence of a formal separation of powers and an overarching constitutional framework[33] in the latter. The dynamics of administrative processes are fundamentally different in MEA compliance systems as states are the creators and enforcers, as well as the subjects and agents, of administrative rules and decisions.[34]

The state-centric settings in MEA compliance systems require a shift in the way that we think about, and analyze, the role of administrative procedures. I propose that, in these settings, desirable proceduralization encompasses: (1) the systematic adoption of administrative procedures, (2) decision making that takes place in procedurally constrained spaces and (3) opportunities for affected states' interests to be voiced and taken into account in decision-making processes. In Chapter 2, I articulate a framework that identifies a range of concrete, institutional practices that provide tangible criteria for evaluating these elements of proceduralization in MEA compliance systems. This framework reflects a synthesis of insights from global administrative law (GAL), international relations, regulatory theory and Third World Approaches to International Law (TWAIL).

There are three elements to this evaluative framework. For the first element, procedural tools for accountability, transparency, participation, reason giving and review from GAL[35] furnish criteria for mapping the degree to which administrative procedures have in fact been incorporated in MEA compliance systems.[36] For the second element, insights from international relations[37] and regulatory theory[38] provide concepts for evaluating the creation of procedurally constrained spaces for decision making.[39] Such constraints place limits on purely political processes and promote administrative decision-making ideals of predictability and fairness.[40] For

33 In IEL, see, eg, D Bodansky, 'Is There an International Environmental Constitution?' (2009) 16(2) *Indiana Journal of Global Legal Studies* 565, 584. Cf Kotze's arguments for the extension of 'the constitutionalist paradigm to the global environmental law and governance domain': L J Kotze, 'Arguing Global Environmental Constitutionalism' (2012) 1(1) *Transnational Environmental Law* 199.

34 Kingsbury, Krisch and Stewart, 'The Emergence of Global Administrative Law', above n 1, 36.

35 GAL can be defined as 'comprising the mechanisms, principles and practices, and supporting social understandings that promote or otherwise affect the accountability of global administrative bodies, in particular by ensuring they meet adequate standards of transparency, participation, reasoned decision, and legality, and by providing effective review of the rules and decisions they make': Kingsbury, Krisch and Stewart, 'The Emergence of Global Administrative Law', above n 1, 17.

36 See Chapter 2, section 3.

37 See, eg, K W Abbott et al, 'The Concept of Legalization' (2000) 54(3) *International Organization* 401, 404; K W Abbott and D Snidal, 'Hard and Soft Law in International Governance' (2000) 54(3) *International Organization* 421.

38 See, eg, N Dubash and B Morgan (eds), *The Rise of the Regulatory State of the South: Infrastructure and Development in Emerging Economies* (Oxford University Press, 2013).

39 See Chapter 2, section 4.

40 D C Esty, 'Good Governance at the Supranational Level: Globalizing Administrative Law' (2006) 115 *Yale Law Journal* 1490, 1529.

the third element of the framework, understandings from GAL[41] and TWAIL[42] inform the identification of indicia of procedural regard and disregard for affected states' interests.[43] As an ideal, procedural regard requires that decision makers systematically consider the interests and views of those affected by their decisions, and justify any decisions that disadvantage particular actors with reference to relevant decisional norms.[44] Of course, in international treaties premised on collective action, the degree of responsiveness to individual states' interests and preferences cannot be unbounded, and ought to be circumscribed by procedural rules, past precedents and standards, and the principles of the regime.[45] This underscores that procedurally constrained decision making and procedural regard are frequently interlinked, and mutually reinforcing.

I argue that proceduralization is desirable in MEA compliance systems for three main reasons. First, as mentioned above, the adoption of administrative procedures has the potential to place constraints on political processes, and reduce the risk that compliance decisions are strongly influenced by differences in power between developed and developing states that have roots in colonialism. If the aims of procedural regard are achieved, there will be opportunities for developing and transition states, which constitute the majority of states affected by compliance decisions,[46] to voice their interests so they can be taken into account in decisions affecting them. If this potential is realized, administrative procedures can contribute to leveling the playing field for less powerful states by providing opportunities for their compliance interests and concerns to be articulated, and considered in decision-making processes.

Second, proceduralized compliance processes should allow for greater interest representation for affected actors, which is a key value associated with due process in international law.[47] This value is critically important in MEA compliance systems, as allowing states to explain the reasons for their non-compliance facilitates decision making that is responsive to affected states' needs and circumstances. There is a wide range of factors that may contribute to a state's non-compliance

41 See, eg, Stewart, 'Remedying Disregard in Global Regulatory Governance', above n 20.

42 See, eg, Chimni, 'Co-option and Resistance', above n 24, 826; B S Chimni, 'Global Administrative Law: Winners and Losers' (Paper presented at the NYU Law School Global Administrative Law Conference, 22–23 April 2005) 11 <http://iilj.org/global_adlaw/documents/ChimniPaper.pdf>.

43 See Chapter 2, section 5.

44 Stewart, 'Remedying Disregard in Global Regulatory Governance', above n 20, 224–225.

45 For a similar point made in the context of discussing legalization, see Abbott and Snidal, 'Hard and Soft Law in International Governance', above n 37, 430–431.

46 See above n 22.

47 Hovell distinguishes between three core values underlying the dialogue enabled by due process in international settings: (1) enhancing the accuracy of decision-making outcomes, (2) providing interest representation for those affected by decisions and (3) promoting public accountability in decision-making: Hovell, *The Power of Process*, above n 32, 63ff. With regard to (2), Hovell's focus is on the interests of affected individuals, while the focus here is on interest representation for affected states.

with their commitments under MEAs, including lack of capacity,[48] inadvertence,[49] conflicting domestic priorities,[50] lack of political will[51] and unforeseen events. Non-compliance processes in MEAs act as 'preventive mechanisms' before more serious problems of environmental damage arise.[52] This function is enhanced by creating forums in which the reasons for states' non-compliance can be heard and understood. Such opportunities signal respect for states' autonomy, and diverse capacities and circumstances. They are also a valuable precursor to tailoring appropriate responses to address non-compliance, which in some cases includes financial and technical support to assist Parties to meet their obligations and achieve long-term compliance. Such responsiveness may, in turn, valuably foster trust and participation in an IEL context premised on state consent, co-operation and goodwill.[53] MEAs address global environmental concerns that by their very nature require collective action to address community interests, underscoring the importance of commitment from all state Parties.[54]

Third, the importance of procedural rigor is heightened in treaties providing for relatively strong measures against non-complying states. In addition to the facilitative measures mentioned above, the Montreal, Kyoto and CITES compliance systems are similar – and diverge from other global MEA compliance systems – as they make provision for intrusive consequences for non-compliance.[55] Specifically, the range of possible responses to non-compliance in the case-study MEAs includes trade sanctions and the deprivation of other rights and privileges under the treaty

48 As Chayes and Handler Chayes note, developing countries' implementation and compliance issues are generally attributable to 'a severe dearth of the requisite scientific, technical, bureaucratic, and financial wherewithal to build effective domestic enforcement systems': Chayes and Handler Chayes, *The New Sovereignty*, above n 9, 14.

49 See, eg, R B Mitchell, 'Compliance Theory: An Overview' in J Cameron, J Werksman and P Roderick (eds), *Improving Compliance with International Environmental Law* (Earthscan, 1996) 3, 11–13.

50 G M Bankobeza, *Ozone Protection: The International Legal Regime* (Eleven International Publishing, 2005) 226.

51 E Couzens, *Whales and Elephants in International Conservation Law and Politics: A Comparative Study* (Routledge, 2014) 128.

52 T Stephens, *International Courts and Environmental Protection* (Cambridge University Press, 2009) 87.

53 Franck argues that adhering to 'right process' promotes states' voluntary compliance with international law: Franck, *Fairness in International Law and Institutions*, above n 15, 26; see also L Tomlinson, *Procedural Justice and the United Nations Framework Convention on Climate Change: Negotiating Fairness* (Springer, 2015) 74.

54 J Brunnée, 'Compliance Control' in G Ulfstein, T Marauhn and A Zimmermann (eds), *Making Treaties Work: Human Rights, Environment and Arms Control* (Cambridge University Press, 2007) 373, 383.

55 J Bulmer, 'Compliance Regimes in Multilateral Environmental Agreements' in J Brunnée, M Doelle and L Rajamani (eds), *Promoting Compliance in an Evolving Climate Regime* (Cambridge University Press, 2012) 55, 71–73; Ulfstein, 'Dispute Resolution, Compliance Control and Enforcement', above n 4, 130.

in question.[56] Ensuring that decision making is not unduly politicized and is informed by a nuanced understanding of the reasons for non-compliance and the effects of the proposed decision upon affected Parties has value in all MEA compliance determinations, and especially those that can result in punitive measures against states.[57]

3 Outline of the book

The evaluative framework for proceduralization introduced in the previous section is elaborated upon in Chapter 2. Thereafter, Chapter 3 canvasses the institutional 'nuts and bolts' of the Montreal, Kyoto and CITES compliance systems, and further develops the rationale for the selection of the case studies. In order to illuminate the impact of administrative procedures on the balance of power between states in a context of North–South inequalities, the focus is on global MEAs, rather than regional MEAs.[58] The selected case studies are three of only seven global MEAs that meet the definition of compliance systems outlined above.[59] Of these seven, the case studies were selected on the basis of being actively utilized,[60] and their relevant similarities and differences in procedural design.[61] In the light of the adoption of the Paris Agreement under the United Nations Framework

56 See Chapter 3, section 5.

57 See also Esty, 'Good Governance at the Supranational Level', above n 40, 1509–1514; Ulfstein, 'Dispute Resolution, Compliance Control and Enforcement', above n 4, 128, 133.

58 Key examples of the latter include the *Convention on Access to Information, Public Participation in Decision-Making and Access to Justice in Environmental Matters*, opened for signature 25 June 1998, 38 ILM 517 (entered into force 30 October 2001), *Convention on Long Range Transboundary Air Pollution*, opened for signature 13 November 1979, 18 ILM 1442 (entered into force 16 March 1983), *Espoo Convention on Environmental Impact Assessment in a Transboundary Context*, opened for signature on 25 February 1991, 1989 UNTS 310 (entered into force 10 September 1997), *Protocol on Water and Health to the 1992 Convention on the Protection and Use of Transboundary Watercourses and International Lakes*, opened for signature 17 June 1999, 2331 UNTS 202 (entered into force 4 August 2005).

59 Rose, 'Compliance with Multilateral Environmental Agreements', above n 3, 10. The other four are *Convention for the Protection of World Cultural and Natural Heritage 1972*, opened for signature 16 November 1972, 11 ILM 1358 (entered into force 17 December 1975) ('World Heritage Convention'), *Basel Convention on the Control of Transboundary Movements of Hazardous Wastes and their Disposal 1989*, opened for signature 22 March 1989, 1673 UNTS 57 (entered into force 5 May 1992) ('Basel Convention'), *Cartagena Protocol on Biosafety to the Convention on Biological Diversity*, opened for signature 15 May 2000, 2226 UNTS 208 (entered into force 5 June 1992) ('Cartagena Protocol'), and *United Nations Framework Convention on Climate Change*, opened for signature 9 May 1992, 1771 UNTS 107 (entered into force 21 March 1994) ('UNFCCC').

60 Thus, those compliance systems that have heard relatively few or no compliance matters, such as the Cartagena Protocol and the UNFCCC's Multilateral Consultative Process, were excluded as case-study candidates.

61 See further Chapter 3, section 5.

Convention on Climate Change in 2015,[62] it is highly unlikely that the Kyoto Protocol will continue to operate past the end of the second commitment period in 2020. Nonetheless, analysis of the Kyoto Protocol's compliance system in Chapter 5 yields valuable insights regarding proceduralized compliance processes in IEL that provide transferable lessons for the future development of the Paris Agreement, which is discussed in Chapter 7.

The evaluative framework articulated in Chapter 2 is applied to the compliance systems of the Montreal, Kyoto and CITES compliance systems, respectively, in Chapters 4–6. The order of the case studies reflects decreasing benefits of administrative procedures in each compliance system, highlighting both the promises and pitfalls of proceduralization in global environmental governance. Each of the case-study chapters is structured similarly with a focus on procedural tools for accountability, transparency, participation, reason giving and review, and evaluation of whether and under what conditions these procedures constrain power politics and facilitate opportunities for affected states' interests to be heard and considered in compliance decision making. To further crystallize the issues explored in this thematic analysis, each case study concludes with a vignette of compliance decision making in practice.

The evaluative process in the case-study chapters is based on an analysis of publicly available primary- and secondary-source materials. It identifies indicative linkages, rather than causal connections, between adopting administrative procedures and disciplining power politics, and enhancing regard for affected interests.[63] The burden of the analysis in each case study is on the extent to which proceduralization places constraints upon power, and promotes procedural rigor and responsiveness to states' compliance interests. It is beyond the scope of this study to systematically or empirically evaluate the impacts of such procedures on environmental outcomes per se,[64] which is an important opportunity for further research.

My main finding is that, under certain conditions, the adoption of administrative procedures limits the influence of states' asymmetric power on compliance decision

62 *Paris Agreement to the United Nations Framework Convention on Climate Change*, opened for signature 12 December 2015 (entered into force 4 November 2016) (in UNFCCC, Report of the Conference of the Parties on its Twenty-First Session, Addendum, UN Doc FCCC/CP/2015/10/Add.1, 29 Jan 2016) ('Paris Agreement').

63 It would be difficult to establish causal connections between the adoption of administrative procedures and regard for affected interests, particularly as this would require empirical research that is able to isolate the impact of administrative procedures from other important contextual variables. Empirical research methods were not pursued here due to, inter alia, compliance decision-makers' confidentiality constraints. For a more detailed methodological discussion, see A Huggins, *The Desirability of Administrative Proceduralisation: Compliance Rules and Decisions in Multilateral Environmental Agreements* (PhD thesis, UNSW, October 2015).

64 Gupta and Mason's work on transparency indicates that there is as yet insufficient evidence to support a conclusion that greater transparency in global environmental governance substantively impacts environmental performance: M Mason and A Gupta, 'Transparency Revisited' in A Gupta and M Mason (eds), *Transparency in Global Environmental Governance: Critical Perspectives* (MIT Press, 2014) 321, 336–337.

making, and promotes consideration of affected states' interests. The Montreal compliance system case study exemplifies a compliance framework that has been designed with an emphasis on understanding the reasons for Parties' non-compliance and supporting non-compliant states' return to compliance. It highlights how conditional funding can be employed as an effective regulatory technique to buttress procedural regard opportunities and to ensure that developing and transition states have the financial and technical means to achieve compliance. The case study on the Kyoto compliance system highlights the promises of procedurally constrained decision-making spaces with extensive due-process guarantees, as well as the missed opportunity for procedural regard if decision-making power nonetheless concentrates in less formal negotiating spaces with thinner procedural safeguards. The CITES compliance system demonstrates the risk that weak and unevenly adopted procedural safeguards allow power politics to operate largely unchecked, evidenced by a trend in which developing country range states are the primary subjects of compliance proceedings and trade suspensions, while developed country export states are not similarly targeted.[65] However, as the values within the regime have shifted away from a narrow focus on a preservationist agenda to a more holistic, pluralist understanding of ways to promote conservation, an environment that is more conducive to procedural regard has emerged.

A synthesis of the case-study analyses reveals three key overarching trends. First, administrative procedures are being adopted, albeit to varying degrees, in these MEA compliance systems. Second, administrative procedures have demonstrated potential to limit political influences on compliance decision making. However, in the absence of comprehensive adoption of administrative procedures, frequently an increase in procedural safeguards in one domain is accompanied by the emergence of, and a migration of regulatory power to, decision-making spaces where such safeguards are weaker. This highlights the power-constraining qualities of administrative procedures, and raises questions about the overall net gains in the absence of systematic adoption. Third, although there is little evidence of administrative procedures being co-opted by powerful states, states' ability to use procedural mechanisms plays an important role in realizing latent procedural regard opportunities. Such insights valuably inform future procedural design choices in MEA compliance systems, as is explored in Chapter 7.

65 On the latter skew in compliance decisions, see also P H Sand, 'Enforcing CITES: The Rise and Fall of Trade Sanctions' (2013) 22(3) *Review of European Community & International Environmental Law* 251, 261.

2 Evaluating proceduralization in MEA compliance systems

1 Introduction

In this chapter, the framework for evaluating the benefits of proceduralization in MEA compliance systems is elaborated upon and contextualized. I argue that regulatory processes within MEA compliance systems represent administrative action, which should be constrained by administrative procedures and mechanisms. Thereafter, the proceduralization framework, which facilitates evaluation of the extent to which the adoption of administrative procedures contributes to procedurally constrained spaces for decision making, and enhances regard for affected states' interests, is explained. Specifically, a set of procedural tools from global administrative law, which provide criteria for mapping the extent of adoption of administrative procedures in MEA compliance systems, is identified. Then, insights from international relations and regulatory theory highlight the desirability of procedurally constrained spaces for decision making as an alternative to unfettered political negotiations, and strict rules-based processes. Finally, indicia of procedural regard and disregard, which furnish criteria for establishing the extent to which administrative procedures enhance decision makers' consideration of the interests of affected states, are canvassed.

2 Administrative action in MEA compliance systems

Theories of global administrative law,[1] global constitutionalism,[2] international institutional law[3] and international public law[4] all seek to advance a public-law

1 See, eg, B Kingsbury, N Krisch and R B Stewart, 'The Emergence of Global Administrative Law' (2005) 68 *Law and Contemporary Problems* 15; N Krisch and B Kingsbury, 'Introduction: Global Governance and Global Administrative Law in the International Legal Order' (2006) 17(1) *The European Journal of International Law* 1, and the other articles in this Symposium issue of the *European Journal of International Law*, 1–278; S Cassese (ed), *Research Handbook on Global Administrative Law* (Edward Elgar, 2016).

2 See, eg, M Kumm, 'The Legitimacy of International Law: A Constitutionalist Framework Analysis' (2004) 15 *European Journal of International Law* 907; J Klabbers, A Peters and G Ulfstein (eds), *The Constitutionalization of International Law* (Oxford University Press, 2009).

3 See, eg, J Klabbers, *An Introduction to International Institutional Law* (Cambridge University Press, 2nd ed, 2009); H G Schermers and N Blokker, *International Institutional Law: Unity within Diversity* (Brill, 5th ed, 2011).

4 A von Bogdandy, M Goldmann and I Vengke, 'From Public International to International Public Law: Translating World Public Opinion into International Public Authority' (2017) 28(1) *European Journal of International Law* 115.

paradigm in international law. Of these, GAL is the most apposite for present purposes given its discrete and 'modest' focus on specific institutional practices of an administrative character.[5] A key insight from GAL is that regulatory rules and decisions in global governance can often be conceptualized as administrative action, which should be circumscribed by administrative mechanisms.[6] GAL scholars argue that global administrative action includes elements of rule making, adjudication and other decision making that fall short of formal treaty making and adjudicative dispute settlement between the Parties.[7] As shall be shown, both these descriptors apply to MEA compliance systems.

First, the primacy of state consent, which traditionally underpins formal treaty processes in international law, is diminishing with respect to 'regime-based law-making processes' in IEL.[8] Traditional consent-based practices are increasingly supplemented by 'law emerging from simplified amendment procedures and secondary decisions of competent treaty bodies'.[9] Simplified amendment procedures allow the relevant Conference of the Parties (COP) to exercise delegated decision-making powers under the aegis of a treaty framework with implications for all Parties.[10] Secondary decisions of treaty bodies, such as compliance bodies, reflect the exercise of formal or de facto delegated authority derived from the superior treaty and reinforced by agreement among the member states.[11] Such practices apply to, inter alia, adjusting Parties' obligations in response to new circumstances, clarifying technical details, supervising implementation and taking action on non-compliance. While the formal legal status of these types of administrative decision making in

5 N Krisch, 'Global Administrative Law and the Constitutional Ambition' in P Dobner and M Loughlin (eds), *The Twilight of Constitutionalism* (Oxford University Press, 2010) 245, 255–259.

6 Kingsbury, Krisch and Stewart, 'The Emergence of Global Administrative Law', above n 1, 28.

7 *Ibid* 17; Krisch, 'Global Administrative Law and the Constitutional Ambition', above n 5, 255.

8 J Brunnée, 'Common Areas, Common Heritage, and Common Concern' in D Bodansky, J Brunnée and E Hey (eds), *The Oxford Handbook of International Environmental Law* (Oxford University Press, 2007) 550, 569; A Peters, 'Compensatory Constitutionalism: The Function and Potential of Fundamental International Norms and Structures' (2006) 19 *Leiden Journal of International Law* 579, 587–589. On the topic of the decay of consent in international law generally, see N Krisch, 'The Decay of Consent: International Law in an Age of Global Public Goods' (2014) 108(1) *American Journal of International Law* 1.

9 T Gehring, 'Treaty-Making and Treaty Evolution' in D Bodansky, J Brunnée and E Hey (eds), *The Oxford Handbook of International Environmental Law* (Oxford University Press, 2007) 467, 469.

10 Often such COP decisions include opt-out provisions for dissenting states. For example, in CITES the COP is responsible for listing and delisting species from Appendices I and II in accordance with Article XV of the Convention, and Parties have the opportunity to opt out by entering reservations against any COP amendments. See further Chapter 6, section 2.1.3.

11 Gehring, 'Treaty-Making and Treaty Evolution', above n 9, 488–495; J Brunnée, 'COPing with Consent: Law-Making under Multilateral Environment Agreements' (2002) 15(1) *Leiden Journal of International Law* 1, 15–31.

MEAs is contested,[12] for a variety of reasons, states tend to commit to and honour such decisions in the practical operation of environmental treaty systems.[13] The importance of administrative constraints on power is heightened in the context of the diminishing influence of state consent as a legal constraint in global environmental governance.

Second, MEA compliance systems do not usually include traditional adjudicative dispute settlement between Parties, and thus do not fit neatly with accounts of the 'judicialization of international politics'.[14] Although MEA compliance systems exist without prejudice to dispute settlement mechanisms,[15] formal independent review of this nature has not yet occurred in practice, and is not a focus of this book. Salient differences between MEA compliance systems and traditional means of settling disputes under treaty law include: the possibility of internal resolution of disputes, the absence of a requirement for consent or an 'injured state', findings of non-compliance that are not legally binding, a recognition of the importance of continued participation for the achievement of common treaty goals and the inclusion of preventive measures such as verification.[16] Accordingly, euphemisms such as 'non-compliance' or 'compliance matters' are preferred over 'breach of treaty'.[17] In sum, therefore, rules and decisions within MEA compliance systems depart from consent-based notions of treaty making, and were consciously designed as an alternative to bilateral dispute settlement, reinforcing the suitability of describing compliance processes as administrative action.

12 Gehring, 'Treaty-Making and Treaty Evolution', above n 9, 492–495. Note that from a GAL perspective, Kingsbury proposes an alternative vision of legality to state consent, which is premised on a 'social fact' conception of law and requirements of 'publicness': B Kingsbury, 'The Concept of "Law" in Global Administrative Law' (2009) 20(1) *European Journal of International Law* 23.

13 Gehring, 'Treaty-Making and Treaty Evolution', above n 9, 492–495; see also J Brunnée, 'International Legal Accountability through the Lens of the Law of State Responsibility' (2005) 36 *Netherlands Yearbook of International Law* 21, 53.

14 See, eg, K J Alter, *The New Terrain of International Law: Courts, Politics, Rights* (Princeton University Press, 2014) 335.

15 For example, the outcomes reached under the Montreal Protocol's non-compliance procedure do not foreclose the possibility of the matter being considered under the Ozone Convention's dispute resolution provisions (Article 11), which operates in parallel with the non-compliance procedure. Similarly, CITES provides for dispute settlement using the Permanent Court of Arbitration in the Hague (Article XVIII). Article 14 of the UNFCCC provides for the settlement of disputes regarding the interpretation or application of the Convention through the International Court of Justice, international arbitration or through a conciliation commission. The application of this provision *mutatis mutandis* is explicitly provided for in Article 19 of the Kyoto Protocol.

16 See, eg, M A Fitzmaurice and C Redgwell, 'Environmental Non-Compliance Procedures and International Law' (2000) 31 *Netherlands Yearbook of International Law* 35, 40–42.

17 M Koskenniemi, 'Breach of Treaty or Non-Compliance? Reflections on the Enforcement of the Montreal Protocol' (1992) 3(1) *Yearbook of International Environmental Law* 123, 128, 145–146.

3 Identifying procedural tools: insights from global administrative law

To what extent is the administrative action in MEA compliance systems circumscribed by administrative procedures and mechanisms? GAL provides a useful descriptive and taxonomic framework to facilitate systematic study of procedures in global governance.[18] In particular, GAL scholarship identifies norms and practices for accountability, transparency, participation, reason giving and review,[19] which inform the identification of procedural tools[20] of an administrative character in the proceduralization framework. These procedural tools are informed by legal principles and practices from many Western domestic legal systems,[21] yet are adapted to the distinctive political and legal context in global environmental governance,[22] which modulates the effects of adopting administrative procedures.

A clear articulation of a set of GAL norms and practices is provided by Kingsbury, Krisch and Stewart in their highly influential 2005 article:

> In our approach, global administrative law effectively covers all the rules and procedures that help ensure the ***accountability*** of global administration, and it focuses in particular on administrative structures, on ***transparency***, on ***participatory*** elements in the administrative procedure, on principles of ***reasoned decisionmaking***, and on ***mechanisms of review***.[23]

Accordingly, mechanisms for accountability, transparency, participation, reason giving and review provide a focal point for this evaluative process. These norms and practices are commonly analyzed in the GAL literature,[24] although they are

18 D H Rached, 'Doomed Aspiration of Pure Instrumentality: Global Administrative Law and Accountability' (2014) 3(3) *Global Constitutionalism* 338, 356–357.

19 These norms and practices were first articulated in: Kingsbury, Krisch and Stewart: 'The Emergence of Global Administrative Law', above n 1, 28.

20 The concept of procedural tools applied here draws inspiration from Esty's 'global administrative law toolbox': D C Esty, 'Good Governance at the Supranational Level: Globalizing Administrative Law' (2006) 115 *Yale Law Journal* 1490, 1524–1536.

21 For a critique of the Western biases that obstruct the search for a 'universal set of administrative law principles', see C Harlow, 'Global Administrative Law: The Quest for Principles and Values' (2006) 17(1) *The European Journal of International Law* 187, 207–214.

22 Kingsbury advocates the necessity of 'a theory of international law responsive to … administrative approaches but adapted to global politics and problems': B Kingsbury, 'Global Environmental Governance as Administration: Implications for International Law' in D Bodansky, J Brunnée and E Hey (eds), *The Oxford Handbook of International Environmental Law* (Oxford University Press, 2007) 63, 72.

23 Kingsbury, Krisch and Stewart, 'The Emergence of Global Administrative Law', above n 1, 28 (emphasis added).

24 *Ibid*; see also a similar articulation in B Kingsbury and L Casini, 'Global Administrative Law Dimensions of International Organizations Law' (2009) 6 *International Organizations Law Review* 319, 325; R B Stewart, 'Remedying Disregard in Global Regulatory Governance: Accountability, Participation, and Responsiveness' (2014) 108 *American Journal of International Law* 211, 244–249, 258–268.

rarely broken down and applied with the degree of granularity that characterizes the following discussion. Although I recognize that there are considerable overlaps, complements and synergies between them,[25] I identify and categorize discrete mechanisms corresponding with each of these five norms. Insights from the IEL literature are interwoven into the following discussion to provide a preliminary indication as to how these administrative mechanisms may manifest in MEA compliance systems.

3.1 Accountability

One of the significant contributions of GAL scholarship is the new insights that it provides into accountability arrangements in global governance. However, despite broad agreement on its importance, understandings of accountability diverge.[26] Contemporary usages of the term generally connote both mechanistic and normative dimensions,[27] which can be conceptualized broadly[28] or narrowly.[29] A broad understanding of accountability treats the term as a 'conceptual umbrella',[30] and it appears to be used in this way by Kingsbury, Krisch and Stewart, who indicate that accountability provides an overarching rubric under which mechanisms of transparency, participation, reasoned decision making, and review sit.[31]

In contrast to these broad definitions, accountability can be conceived of as a discrete mechanism alongside the other procedural tools from GAL.[32] Stewart proposes a narrow definition of accountability mechanisms characterized by three structural elements:

> (1) a specified accounter, who is subject to being called to provide account for his conduct; (2) a specified account holder who can require the accounter

25 R B Stewart, 'Accountability, Participation, and the Problem of Disregard in Global Regulatory Governance' (2008) 35–36 <www.iilj.org/courses/documents/2008Colloquium.Session4.Stewart.pdf>.

26 J Koppell, 'Pathologies of Accountability: ICANN and the Challenge of "Multiple Accountability Disorder"' (2005) 65(1) *Public Administration Review* 94, 94.

27 C Harlow, 'Accountability as a Value for Global Governance and Global Administrative Law' in G Anthony, J-B Auby, J Morison and T Zwart (eds), *Values in Global Administrative Law* (Hart Publishing, 2011) 173, 174–182.

28 See, eg, R W Grant and R O Keohane, 'Accountability and Abuses of Power in World Politics' (2005) 99(1) *American Political Science Review* 29, 36; and J L Mashaw, 'Structuring a "Dense Complexity": Accountability and the Project of Administrative Law' (2005) 5(1) *Issues in Legal Scholarship* 1, 27.

29 See, eg, Stewart, 'Remedying Disregard in Global Regulatory Governance', above n 24, 244–255; M Bovens, 'Analysing and Assessing Accountability: A Conceptual Framework' (2007) 13 *European Law Journal* 447, 449–450.

30 Bovens, 'Analysing and Assessing Accountability', above n 29, 449.

31 Kingsbury, Krisch and Stewart, 'The Emergence of Global Administrative Law', above n 1, 28.

32 Stewart, 'Remedying Disregard in Global Regulatory Governance', above n 24, 252–256; Rached, 'Doomed Aspiration of Pure Instrumentality', above n 18, 361 n 78.

> to render account; and (3) the ability and authority of the account holder to impose sanctions or other remedies for deficient performance.[33]

Stewart argues that accountability mechanisms should not be conflated with other types of measures, including those promoting transparency, participation and reason giving – although such measures may facilitate accountability, they are not in themselves accountability mechanisms.[34] For the purposes of identifying tangible criteria against which practices in MEA compliance systems can be assessed, Stewart's narrower definition of accountability is preferable to more expansive definitions. Furthermore, it has prima facie relevance to compliance systems in MEAs, a key objective of which is for states to account for their compliance with their multilateral environmental commitments, and for consequences to be considered and imposed in the event of non-compliance.

As Kingsbury notes, on a functional level, accountability can be analyzed in terms of its components: 'accountability of whom; to whom; for what; and by what means'.[35] These four components provide a further guide to analyzing accountability arrangements in MEA compliance systems. In relation to the first two, there are two strands of inquiry that can be pursued. The first sees individual states as administrative 'agents' and 'subjects' that are ultimately accountable to the Parties for compliance with their treaty obligations.[36] This understanding is highly apt in the context of the administrative action occurring within MEA compliance systems, in which states are subject to the administrative rules and decisions they have collectively created. The second strand sees the decision makers within a compliance system as administrative agents, who are then potentially accountable for their decisions to internal or external oversight bodies. Both of these types of analysis require the 'accounter' and 'account holder' to be identifiable. This necessitates clarity around the identity of the decision maker and steps in the decision-making process. This can be contrasted with 'black box processes', in which the identity of the decision maker is not revealed, and the processes by which decisions are reached occur behind closed doors. [37]

33 Stewart, 'Remedying Disregard in Global Regulatory Governance', above n 24, 253. A similar definition of accountability can be found in J Black, 'Constructing and Contesting Legitimacy and Accountability in Polycentric Regulatory Regimes' (2008) 2 *Regulation and Governance* 136, 150 (defining accountability in terms of the ways in which one actor 'gives account and another [actor] has the power or authority to impose consequences as a result'). According to Stewart, these three structural elements are common to five basic categories of accountability mechanisms: electoral, hierarchical, supervisory, fiscal and legal accountability: *ibid* 245.

34 Stewart, 'Remedying Disregard in Global Regulatory Governance', above n 24, 252–253.

35 Kingsbury, 'Global Environmental Governance as Administration', above n 22, 66. See also C Scott, 'Accountability in the Regulatory State' (2000) 27(1) *Journal of Law and Society* 38, 41–42.

36 Kingsbury, Krisch and Stewart, 'The Emergence of Global Administrative Law', above n 1, 36; J Waldron, 'Are Sovereigns Entitled to the Benefit of the International Rule of Law?' (2011) 22 *European Journal of International Law* 315, 327–337.

37 Esty, 'Good Governance at the Supranational Level', above n 20, 1528–1529.

In relation to Kingsbury's third component, analyzing accountability requires clarity about the standards to which an accounter is being held to account. For example, in MEA compliance systems, are states accountable for compliance with agreed rules with pre-determined consequences for non-compliance? In international institutions such as MEAs, these standards may be derived from treaty obligations, and promote predictability and legality in accordance with the principles of *lex scripta, lex certa lex previa* and *lex stricta*[38] – that is, law that is antecedent, written and clear. Thus, in some situations, the assurance of legality will be interlinked with accountability.[39] However, as shall be shown, in other instances states are held to account for obligations they have negotiated with compliance decision makers that represent a departure from strict adherence to treaty obligations, but are a form of accountability nonetheless.[40]

Regarding Kingsbury's fourth component, there are numerous administrative procedures that serve as means for promoting accountability ends. Inspections and audits are examples of accountability mechanisms that have prima facie relevance to MEA compliance systems, and that can apply to both states and compliance decision makers.[41] Variations on these processes through verification and review of states' compliance with their multilateral environmental commitments are becoming increasingly common[42] – indeed, they often provide the first step in MEA non-compliance processes. Independent audits and inspections of the internal operations of MEA compliance systems, including both decision-making practices and outcomes, can further enhance accountability.[43]

In sum, an important focus of analyzing accountability in this book is on whether there are procedures in place through which state Parties are called to account for, and may be subject to consequences for, non-compliance with their MEA treaty obligations. This entails examining whether there are clear processes to be followed in holding states to account for their international environmental commitments, which countries are called to account before compliance decision-making bodies, which actors make decisions in these bodies and the extent to which the consequences for non-compliance recommended by such bodies are in accordance with relevant decisional norms.

38 See S Borras, 'Comparative Analysis of Selected Compliance Procedures under Multilateral Environmental Agreements' in S Maljean-Dubois and L Rajamani (eds), *Implementation of International Environmental Law* (Martinus Nijhoff Publishers, 2011) 319, 345–346.
39 Kingsbury, Krisch and Stewart, 'The Emergence of Global Administrative Law', above n 1, 29.
40 See, eg, Chapter 4, section 2.
41 Esty, 'Good Governance at the Supranational Level', above n 20, 1525–1526. See also C Harlow, *Accountability in the European Union* (Oxford University Press, 2002) Ch 5.
42 See, eg, K Raustiala, *Reporting and Review Institutions in 10 Multilateral Environmental Agreements* (UNEP, 2001) <www.peacepalacelibrary.nl/ebooks/files/C08-0025-Raustiala-Reporting.pdf>.
43 Esty, 'Good Governance at the Supranational Level', above n 20, 1526.

3.2 Transparency

Transparency is arguably an emerging norm in global environmental governance.[44] It can be conceptualized as 'a governance of *information*, including demands for active transparency and access to information, but also demands for confidentiality and privacy, and for legal or political controls on the gathering and use of policy-shaping information'.[45] Despite their apparent pervasiveness and ease of implementation, Fisher highlights that transparency mechanisms are in fact complex, shaped by substantive choices and have far-reaching operational consequences.[46] The following analysis is alert to a range of issues that are pertinent to evaluating transparency, including internal and external transparency issues, active and passive information transmission, what is and is not made transparent as well as the quality of the transparent information.

A distinction can be drawn between internal and external transparency. With internal transparency, only regime management and members are privy to information gathered through monitoring processes. External transparency occurs where such information is more widely disseminated, including through availability to the public.[47] Internal transparency can be promoted through exchange of data between Parties[48] – in MEA compliance systems, this may particularly apply to the provision of baseline data and performance information, as well as the exchange of scientific, technical and economic information,[49] and openness regarding compliance processes and decision making. External transparency is most often facilitated by public access to hearings and information, including through Internet forums.[50] Furthermore, some international institutions' transparency policies permit non-state actors' requests for access to information that is otherwise not publicly available.[51] Generally, international institutions only make available materials and

44 See, eg, A Gupta and M Mason (eds), *Transparency in Global Environmental Governance: Critical Perspectives* (MIT Press, 2014); D B Hunter, 'The Emerging Norm of Transparency in International Environmental Governance' in P Ala'i and R G Vaughn (eds), *Research Handbook on Transparency* (Edward Elgar Publishing, 2014) 343.

45 Kingsbury and Casini, 'Global Administrative Law Dimensions of International Organizations Law', above n 24, 325 (emphasis in original).

46 E Fisher, 'Transparency and Administrative Law: A Critical Evaluation' (2010) 63 *Current Legal Problems* 272.

47 R B Stewart, M Oppenheimer and B Rudyk, 'Building Blocks for Global Climate Protection' (2013) 32 *Stanford Environmental Law Journal* 341, 385–386.

48 For example, this is evident in relation to states' national reports on their compliance with the relevant multilateral environmental commitments under each treaty.

49 J Brunnée and E Hey, 'Transparency and International Environmental Institutions' in A Bianchi and A Peters (eds), *Transparency in International Law* (Cambridge University Press, 2013) 23, 38–39.

50 D C Esty, 'Environmental Protection in the Information Age' (2004) 79 *New York University Law Review* 115, 160–161, 167–170.

51 M Donaldson and B Kingsbury, 'The Adoption of Transparency Policies in Global Governance Institutions: Justifications, Effects, and Implications' (2013) 9 *Annual Review of Law and Social Science* 119, 122–124.

documents they have already decided to produce, meaning that the information that is made transparent will be shaped by the institution's view of itself, its mandate and its work in the world.[52]

One of the particular challenges for transparency in treaty-based regimes is the traditional diplomatic norm of confidentiality in negotiations. A focus on the rhetoric and institutionalization of transparency can serve to obscure the aspects of regimes that continue to be governed, implicitly or explicitly, by routine diplomatic and bureaucratic secrecy. In this regard, it is important to bear in mind that MEA compliance systems have an adjudicatory function, meaning that some confidentiality constraints in these forums may well be appropriate.[53] A related challenge is a shared understanding that the resolution of certain technical issues is the appropriate domain of experts,[54] which may constitute a further justification for secrecy.

The quality of information is also a salient factor in evaluating transparency. The types of information to which transparency policies apply include 'decisions, policy statements, reports, various kinds of internal documents, minutes or transcripts of proceedings, [and] organized data'.[55] However, rather than simply making an undigested morass of details, much of which may be technically complex, publicly available, Buchanan and Keohane reinforce the importance of the information having an '*epistemic-deliberative* quality'. This means that the material provided is integrated and presented in a way that facilitates public accessibility, engagement and evaluation of substantive issues.[56] There is, however, a risk that selective editing or over-simplification[57] to produce publicly digestible information could subtly blur into 'governance by information', whereby information is 'sugar-coated' or in other ways selected or moulded to influence public perceptions.[58] The volume of information provided in MEA compliance systems, the languages in which it is available and its technical density all impact its 'epistemic-deliberative quality',[59] and will be unpacked in the following analysis.

52 *Ibid* 121, 133.
53 *Ibid* 120–121.
54 Stewart, 'Accountability, Participation, and the Problem of Disregard in Global Regulatory Governance', above n 25, 68.
55 *Ibid* 32 fn 124.
56 A Buchanan and R O Keohane, 'The Legitimacy of Global Governance Institutions' (2006) 20(4) *Ethics & International Affairs* 405, 425–426 (emphasis in original).
57 A Bianchi, 'On Power and Illusion: The Concept of Transparency in International Law' in A Bianchi and A Peters (eds), *Transparency in International Law* (Cambridge University Press, 2013) 1, 16.
58 A von Bogdandy and M Goldmann, 'Taming and Framing Indicators: A Legal Reconstruction of the OECD's Programme for International Student Assessment' in K E Davis, A Fisher, B Kingsbury and S Engle Merry (eds), *Governance by Indicators: Global Power Through Quantification and Rankings* (Oxford University Press, 2012) 52, 53–54. See also R B Mitchell, 'Transparency for Governance: The Mechanisms and Effectiveness of Disclosure-Based and Education-Based Transparency Politics' (2011) 70 *Ecological Economics* 1882.
59 Buchanan and Keohane, 'The Legitimacy of Global Governance Institutions', above n 56, 425–426.

3.3 Participation

Avenues for participation allow interested stakeholders to actively engage with the information that has been made transparent through the above-mentioned types of processes and, in some instances, to contribute to compliance deliberations.[60] According to Ebbesson, public participation in international environmental decision making encompasses activities and procedures by which 'individuals and groups can:

- increase their general awareness;
- express their views;
- influence and engage themselves in law- and policy-making;
- affect decision-making concerning specific activities, plans, and programmes;
- trigger reviews of administrative decisions; and/or
- have existing laws enforced'.[61]

Although traditional discourses regarding participation in IEL related to states,[62] in contemporary MEA compliance systems, this norm encompasses states' participation as well as public participation by non-state actors, including non-governmental organizations (NGOs) and other civil-society representatives.[63]

In assessing opportunities for public engagement, Stewart's distinction between non-decisional, juridical and decisional participation rights is pertinent.[64] 'Non-decisional participation practices' include opportunities for consultation or comment and access to meetings where pending decisions are being considered. In contrast, 'juridical participation rights' pertain to an actor's rights to bring an action or present evidence or argument in relation to an administrative decision that affects them, and 'decisional participation rights' come with attendant voting and decision-making authority.[65] The conflation of these three types of participation downplays significant differences in the roles and authority available to participants.

60 Kingsbury, 'Global Environmental Governance as Administration', above n 22, 68. Principle 10 of the Rio Declaration reflects a classic articulation of public participation principles in a soft law instrument: United Nations General Assembly, Rio Declaration on Environment and Development (Annex I), A/CONF.151/26 (vol 1), 14 Jun 1992, reprinted in 31 ILM 874.

61 J Ebbesson, 'Public Participation' in D Bodansky, J Brunnée and E Hey (eds), *The Oxford Handbook of International Environmental Law* (Oxford University Press, 2007) 681, 683.

62 D Bodansky, 'Legitimacy' in D Bodansky, J Brunnée and E Hey (eds), *The Oxford Handbook of International Environmental Law* (Oxford University Press, 2007) 704, 717.

63 See, eg, S Marsden, 'Public Participation in Transboundary Environmental Impact Assessment: Closing the Gap between International and Public Law' in B Jessup and K Rubenstein (eds), *Environmental Discourses in Public and International Law* (Cambridge University Press, 2012) 238, 238–239; K Raustiala, 'NGOs in International Treaty-Making' in D B Hollis (ed), *The Oxford Guide to Treaties* (Oxford University Press, 2011) 150, 170–171.

64 Stewart, 'Accountability, Participation, and the Problem of Disregard in Global Regulatory Governance', above n 25, 8, 34–35.

65 *Ibid.* For the purposes of this book, 'juridical' refers to the administration of the law, and can be distinguished from 'judicial', which refers to court proceedings involving judges.

The following analysis distinguishes between these three variants of participation. States' participation options in MEA compliance systems, evidence of which is collated through examining membership requirements for decision-making bodies, voting rights and other formal mechanisms for input into decision making, are likely to encompass all three types of participation rights. In terms of promoting regard for affected states' interests, juridical participation rights are particularly salient. In the light of the sensitivity of non-compliance information involving sovereign states, non-state actors' participation options are expected to be relatively limited.[66]

3.4 Reason giving

Providing reasons for reaching certain types of decisions in global governance can help actors who have the authority to impose decisions to allay concerns about the legitimacy of the relevant decision-making processes.[67] As Esty succinctly observes, '[t]he rationality of a policy choice can best be evaluated when it is written down, explained, and published'.[68] The clear articulation of reasons for a decision promotes the similar treatment of analogous cases, thereby contributing to broader administrative decision-making ideals of fairness and predictability.[69] Where compliance decisions take into account country-specific factors, reason giving also allows the rationale for the decision to be understood. Such clarity means that Parties that are adversely affected can critique, and possibly contest, the basis for the decision.[70]

A number of different types of reasons can be distilled from Shapiro's comparison of American and European reason-giving requirements,[71] which prima facie provide suitable indicia for evaluating compliance decision making in MEAs. The four types are: (1) the record of the final decision only, (2) 'giving reasons of law', which are then potentially reviewable for errors, (3) 'giving reasons of fact', which opens up the possibility of procedural and also substantive review and (4) 'dialogue' in respect of competing considerations, which are again potentially reviewable on both procedural and substantive grounds.[72] These four variations differ in terms of the complexity of reasons given, and therefore the extent to

66 Ebbesson, 'Public Participation', above n 61, 683; J Brunnée, 'Compliance Control' in G Ulfstein, T Marauhn and A Zimmermann (eds), *Making Treaties Work: Human Rights, Environment and Arms Control* (Cambridge University Press, 2007) 373, 386–387.

67 Kingsbury, 'The Concept of "Law" in Global Administrative Law', above n 12, 47. For example, the Aarhus Convention and some World Trade Organization Agreements contain requirements for reason giving: Stewart, 'Accountability, Participation, and the Problem of Disregard in Global Regulatory Governance', above n 25, 35.

68 Esty, 'Good Governance at the Supranational Level', above n 20, 1529.

69 *Ibid.*

70 Stewart, 'Remedying Disregard in Global Regulatory Governance', above n 24, 265.

71 M Shapiro, 'The Giving Reasons Requirement' (1992) *The University of Chicago Legal Forum* 179.

72 *Ibid* 182–187, 201–210.

which they are amenable to review by allowing the appellate decision maker to retrace the original administrator's decision-making process.[73] The following evaluative process is attuned to rules pertaining to reason giving, as well as the extent, types and quality of reasons provided.

3.5 Review

Review mechanisms and appeal processes are common and widely accepted mechanisms of domestic administrative law in both common-law and civil-law traditions.[74] The issue of independent review of the decisions of intergovernmental institutions' actions by external courts such as the International Court of Justice or by other independent judicial or quasi-judicial bodies is becoming more frequent,[75] yet remains contentious.[76] Ulfstein suggests that judicial review for international institutions should ensure that the decision-making process fulfilled relevant procedural requirements, but also that the decision was within the substantive mandate of the relevant body, and accords with general international law.[77] While in theory a Party that has been subject to non-compliance proceedings in a global MEA may still seek an independent opinion before an arbitral tribunal or an international court,[78] formal review of this nature has not yet occurred in practice.

Nonetheless, *internal* review and appeal mechanisms that allow relevant actors to receive a second opinion[79] may exist within treaty regimes. Examples of such

73 M Shapiro, *The Supreme Court and Administrative Agencies* (The Free Press, 1968).

74 See, eg, F Bignami, 'From Expert Administration to Accountability Network: A New Paradigm for Comparative Administrative Law' (2011) 59 *American Journal of Comparative Law* 859, 872 n 32; Kingsbury, Krisch and Stewart, 'The Emergence of Global Administrative Law', above n 1, 39.

75 On the issue of judicial review by the International Court of Justice, particularly in relation to decisions of the UN Security Council, see, eg, E de Wet, 'Judicial Review as an Emerging Principle of Law and its Implications for the International Court of Justice' (2000) 47(2) *Netherlands International Law Review* 181; J Crawford, 'Marbury v. Madison at the International Level' (2004) 36 *George Washington International Law Review* 505.

76 Kingsbury, Krisch and Stewart, 'The Emergence of Global Administrative Law', above n 1, 40; R Mackenzie, 'The Role of Dispute Settlement in the Climate Regime' in J Brunnée, M Doelle and L Rajamani (eds), *Promoting Compliance in an Evolving Climate Regime* (Cambridge University Press, 2012) 395, 413–415.

77 G Ulfstein, 'Institutions and Competences' in J Klabbers, A Peters and G Ulfstein (eds), *The Constitutionalization of International Law* (Oxford University Press, 2009) 45, 67.

78 P Sands, 'Non-Compliance and Dispute Settlement' in U Beyerlin, P-T Stoll and W Rudiger (eds), *Ensuring Compliance with Multilateral Environmental Agreements: A Dialogue Between Practitioners and Academia* (Brill Martinus Nijhoff Publishers, 2006) 353, 356–358.

79 Esty, 'Good Governance at the Supranational Level', above n 20, 1536.

mechanisms, allowing appeals on procedural[80] and substantive grounds,[81] can be found in various pockets of international law. In addition to internal appeals avenues, this analysis also pays attention to instances in which a body that is higher in the hierarchy of a compliance system exercises oversight over, and may even overrule or substitute, a subsidiary body's compliance decision. Such oversight may occur even if a state Party has not formally appealed the decision. While such avenues are 'less formal, visible, and legally structured' than review by independent arbitral bodies or courts,[82] they ensure that compliance decisions by subsidiary bodies can be reconsidered as required.

Issues of interest in terms of analyzing internal appeal and review mechanisms in MEA compliance systems include: the extent to which such avenues exist, which actors are able and enabled to utilize them, the grounds upon which appeals may be made (whether due to alleged procedural error, or errors of fact or law) and whether a successful appeal results in remission for reconsideration by the primary decision maker, or the original decision being overruled and substituted with a new decision. In terms of oversight of a subsidiary body's compliance decision making in the absence of an appeal, key additional questions include which bodies are authorized to review decisions, and how the decision comes to the attention of the superior body.

Thus, GAL norms and practices pertaining to accountability, transparency, participation, reason giving and review inform fine-grained criteria for evaluating the extent of adoption of administrative procedures in MEA compliance systems. In order to evaluate if and how administrative procedures shape and constrain global administrative action, a simultaneous examination of decision-making practices is required.

4 Procedurally constrained spaces for decision making: insights from international relations and regulatory theory

There appears to be a trend toward the adoption of administrative procedures in MEA compliance systems,[83] although further inquiry using the criteria mentioned

80 An example of this is the annulment procedure in the Convention on the Settlement of Investment Disputes: *Convention on the Settlement of Investment Disputes between States and Nationals of Other States*, opened for signature 18 March 1965, 575 UNTS 160 (entered into force 14 October 1966).

81 For example, the appeals procedures in the World Trade Organization allow appeals on the grounds of due process and alleged errors of law: Mackenzie, 'The Role of Dispute Settlement in the Climate Regime', above n 76, 411–412.

82 Stewart, 'Remedying Disregard in Global Regulatory Governance', above n 24, 268.

83 See, eg, A Tanzi and C Pitea, 'Non-Compliance Mechanisms: Lessons Learned and the Way Forward' in T Treves et al (eds), *Non-Compliance Procedures and Mechanisms and the Effectiveness of International Environmental Agreements* (TMC Asser Press, 2009) 569, 580; and K N Scott, 'Non-Compliance Procedures and Dispute Resolution Mechanisms under International Environmental Agreements' in D French, M Saul and N D White (eds),

above is desirable to systematically analyze the extent of this phenomenon. However, there is considerable uncertainty regarding the roles played by these procedures. First, administrative procedures may circumscribe political influences in MEA compliance systems, yet such constraints may be put under strain by the continuing traction of traditional, political forms of decision making characterized by negotiation, diplomacy and co-operation.[84] Second, administrative procedures have potential to promote regard for affected interests, yet concerns have been raised about whether they are vulnerable to being co-opted by, and reinforcing the priorities of, the powerful.[85] These debates inform the remaining two elements of the proceduralization framework.

The first of these debates hinges on the ability of administrative procedures to limit the influence of political power and expediency on decision making.[86] Some commentators advocate for procedural constraints on politics in MEA compliance procedures. For example, both Koskenniemi and Hey recognize the benefits of formalizing compliance procedures and incorporating procedural safeguards.[87] However, there is considerable uncertainty regarding the extent to which this aspiration translates into practice in MEA compliance systems. Despite preferring greater reliance on formalism, procedural safeguards and the rule of law, Klabbers observes that compliance processes in IEL are in practice 'subject to negotiations'.[88] Other scholars, too, draw attention to the pragmatic realities of political and flexible approaches to compliance in international settings.[89] Chayes and Handler

International Law and Dispute Settlement: New Problems and Techniques (Hart Publishing, 2010) 225, 230.

84 L Rajamani and S Maljean-Dubois, 'The Present State of Research Carried Out by the English-speaking and French-speaking Sections – Report of the Directors of Studies' in S Maljean-Dubois and L Rajamani (eds), *Implementation of International Environmental Law* (Martinus Nijhoff Publishers, 2011) 107, 133–134; D Bodansky, *The Art and Craft of International Environmental Law* (Harvard University Press, 2009) 248; M Montini, 'Procedural Guarantees in Non-Compliance Mechanisms' in T Treves et al (eds), *Non-Compliance Procedures and Mechanisms and the Effectiveness of International Environmental Agreements* (TMC Asser Press, 2009) 389, 390–392.

85 See, eg, B S Chimni, 'Co-option and Resistance: Two Faces of Global Administrative Law' (2005) 37 *International Law and Politics* 799, 803.

86 See, eg, Stewart, who argues that global administrative procedures 'operate to limit decisions on the basis of power and expediency': Stewart, 'Remedying Disregard in Global Regulatory Governance', above n 24, 220. Similarly, Chesterman notes that the goals of GAL include, but go beyond, 'constraining decision-makers': S Chesterman, 'Globalisation and Public Law: A Global Administrative Law?' in J Farrall and K Rubenstein (eds), *Sanctions, Accountability and Governance in a Globalised World* (Cambridge University Press, 2009) 75, 87–88.

87 Koskenniemi, 'Breach of Treaty or Non-Compliance?', above n 17, 147; E Hey, 'International Institutions' in D Bodansky, J Brunnée and E Hey (eds), *The Oxford Handbook of International Environmental Law* (Oxford University Press, 2007) 749, 767.

88 J Klabbers, 'Compliance Procedures' in D Bodansky, J Brunnée and E Hey (eds), *The Oxford Handbook of International Environmental Law* (Oxford University Press, 2007) 995, 1001, 1007–1009.

89 See, eg, N Goeteyn and F Maes, 'Compliance Mechanisms in Multilateral Environmental Agreements: An Effective Way to Improve Compliance?' (2011) 10 *Chinese Journal of*

Chayes are perhaps the strongest advocates for the *desirability* of a 'co-operative, problem-solving approach' to promoting compliance with international agreements such as MEAs.[90] According to this view, political modes of decision making are an inescapable reality of compliance in international law, which can be harnessed to promote compliance outcomes.

It is clear, therefore, that in IEL, as in other regulatory contexts, there is a tension between the administrative apparatus and the political context.[91] Gauging the extent of this dissonance thus becomes critical. As indicated above, the starting point for this analysis is mapping the presence of practices of accountability, transparency, participation, reason giving and review in MEA compliance systems. The second step involves evaluating the extent to which such mechanisms contribute to the creation of procedurally constrained spaces[92] for compliance deliberations. This concept acknowledges the inevitability of politics in compliance proceedings in IEL,[93] as well as the desirability of placing some limits on political negotiations, especially in the context of significant differences in bargaining power between states. Such constraints are an important antecedent to regard for affected states' interests.

A spectrum provides a valuable heuristic device for gauging the extent to which regulatory decision making is primarily political, or constrained by administrative rules and procedures. From an international relations perspective, Abbott et al specify that one dimension of the legalization of world politics is the delegation of legal authority, ranging along a spectrum from political bargaining and diplomacy to binding, third-party dispute settlement mechanisms.[94] As noted above, MEA compliance systems developed in response to the limitations of adjudicative dispute settlement, and thus fall short of legalistic approaches to implementing international agreements. However, they are also proceduralized domains of governance, which

International Law 791, 826; Bodansky, *The Art and Craft of International Environmental Law*, above n 84, 248.

90 A Chayes and A Handler Chayes, *The New Sovereignty: Compliance with International Regulatory Agreements* (Harvard University Press, 1998) 3.

91 See, eg, N Dubash and B Morgan, 'The Embedded Regulatory State: Between Rules and Deals' in N Dubash and B Morgan (eds), *The Rise of the Regulatory State of the South: Infrastructure and Development in Emerging Economies* (Oxford University Press, 2013) 279, 279–296.

92 This concept is informed by the work of Dubash and Morgan: *ibid* 289.

93 As Abbott and Snidal note, 'Politics permeates international law and limits its autonomy ... The deep connection of law and politics is most apparent in the area of delegation', an example of which is COPs delegating authority to MEA compliance systems: K W Abbott and D Snidal, 'Hard and Soft Law in International Governance' (2000) 54(3) *International Organization* 421, 455.

94 The other two dimensions are obligation, which ranges from binding rules to expressly non-legal norms, and precision, which ranges from precise, highly elaborate rules to vague principles: K W Abbott et al, 'The Concept of Legalization' (2000) 54(3) *International Organization* 401, 404, 408–418.

are anticipated to place some constraints on political bargaining.[95] As such, compliance procedures can be seen as a type of 'soft law', which can be distinguished from 'hard law' on the one hand, and purely political arrangements on the other.[96] If such constraints are ineffectual, compliance may be subject to political negotiations without reference to a principled justification.[97] Where this occurs, the political influences on decision making can include power politics and/or narrow, special-interest politics serving private ends,[98] both of which are highly undesirable in compliance determinations.

Another useful spectrum is proposed by Dubash and Morgan in their work examining the interplay between proceduralization and the politics of negotiation in domestic regulatory settings. These authors propose a spectrum 'between rules and deals', and argue that domestic governance contexts can be located at different points along this continuum.[99] The authors propose this spectrum in relation to the emergent regulatory states of the global South. For the purposes of the proceduralization framework, Dubash and Morgan's spectrum needs to be tailored to an IEL context that differs significantly from domestic governance settings. Notably, as there is no formal separation of powers or constitutional framework in IEL,[100] a formal division between administrative and political functions is not necessarily possible nor desirable. Nevertheless, I contend that in order for administrative procedures to have a beneficial effect, it is important that they place some fetters upon purely power-based processes.

Accordingly, an adapted spectrum is required to evaluate whether increasing reliance on administrative procedures in MEA compliance systems is placing constraints upon political modes of decision making, or whether, despite the incorporation of some administrative procedures, political processes remain largely unchecked. Specifically, synthesizing the above insights from international relations and regulatory theory, I propose a spectrum between political decision-making modes, characterized by negotiation, diplomacy and co-operation, and administrative decision-making modes, reflecting autonomous, impartial and rules-based decision making. The ideal characteristics of both styles of decision making are depicted in Figure 1:

95 *Ibid* 416.

96 Abbott and Snidal, 'Hard and Soft Law in International Governance', above n 93, 422. The authors define hard law as 'legally binding obligations that are precise (or can be made precise through adjudication or the issuance of detailed regulations) and that delegate authority for interpreting and implementing the law'. In contrast, soft law 'begins once legal arrangements are weakened along one or more of the dimensions of obligation, precision, and delegation': *ibid* 421–422.

97 Abbott et al, 'The Concept of Legalization', above n 94, 415.

98 Dubash and Morgan, 'Between Rules and Deals', above n 91, 291.

99 *Ibid* 279–296.

100 D Bodansky, 'Is there an International Environmental Constitution?' (2009) 16(2) *Indiana Journal of Global Legal Studies* 565, 584.

Figure 1: A spectrum between political and administrative decision-making modes

It is recognized that, in practice, all decision making in global regulatory bodies encompasses elements of these two ideal types of decision making. Nonetheless, the 'basic *posture*' of decision making is likely to align with one type more than the other.[101] The concept of a spectrum between political and administrative decision-making modes provides a useful analytical device because it acknowledges the tensions and differences between these two styles of decision making, and yet accommodates the reality of interactions between them. Importantly, if administrative procedures promote the creation of autonomous, impartial and rules-based spaces for compliance deliberations, they will place some constraints on political negotiations.

It is important to note that decision making that is autonomous and rules-based does not necessarily deny the appropriate role of administrative discretion in some instances. That is, the adoption of procedural tools of an administrative character can constrain political influences in a way that either limits or preserves discretionary leeway for decision makers.[102] Moreover, administrative decision making does not need to be undertaken by experts. There are two competing models regarding the membership of MEA compliance bodies: a body that is composed of representatives of a small number of Parties to the treaty, and a body that consists of ostensibly independent, technical experts. While in earlier MEAs, such as CITES and the Montreal Protocol, a preference for political compliance bodies was evident, in more recent MEAs, such as the Kyoto Protocol, the choice of an independent body of experts is increasingly prevalent.[103] Both types of bodies can engage in administrative decision making. As with state representatives, where

101 P Nonet and P Selznick, *Toward Responsive Law: Law & Society in Transition* (Transaction Publishers, 2001) 17 (emphasis in original).

102 Fisher's conceptual work on administrative constitutionalism identifies two paradigms of public administration – the rational-instrumental and the deliberative-constitutive paradigms – which, among other important differences, reflect a divergent understanding of the appropriate role for administrative discretion in public administration: E Fisher, *Risk Regulation and Administrative Constitutionalism* (Hart Publishing, 2010) 26–35. Transposing these insights to an IEL context and the spectrum described above, I suggest that both paradigms reflect administrative decision-making modes that can occur within procedurally constrained zones for decision making.

103 A Fodella, 'Structural and Institutional Aspects of Non-Compliance Mechanisms' in T Treves et al (eds), *Non-Compliance Procedures and Mechanisms and the Effectiveness of International Environmental Agreements* (TMC Asser Press, 2009) 355, 359–361.

experts are involved it is inevitable that there will be political dimensions to decision making. Yet, in order to align with ideals of administrative decision making, there should be safeguards in place to help ensure that decision makers are not significantly skewed by personal or political preferences.[104] Hence, procedurally constrained spaces for decision making may exist with or without discretionary leeway for administrative decision makers, and with or without input from experts.

5 Indicia of procedural regard: insights from global administrative law and third world approaches to international law

In addition to the uncertainty regarding the role that procedures play in constraining politics, there are mixed views on whether administrative procedures facilitate regard for affected interests, including the interests of less powerful states, or simply entrench the dominance of the powerful. This uncertainty highlights the need for evaluative criteria to illuminate whether the adoption of administrative procedures is, in fact, regard-enhancing as the third element of the proceduralization framework.

Stewart, a leading proponent of GAL, is a strong advocate for the potential of administrative procedures to promote procedural regard for weaker actors' interests.[105] He defines procedural regard for individuals and groups as follows:

> As an ideal, regard requires that the decision-maker review available information about the effects of proposed decisions on the various groups, individuals, interests, and concerns entitled to consideration; weighs the benefits for and burdens on them of alternatives; and determines that decisions that impose disadvantage or harm on some affected groups and individuals are justified by relevant decisional norms.[106]

In contrast, procedural disregard is evinced by decision makers' failure or refusal to do these things. Procedural regard can also be contrasted with substantive regard, which implies decision-making *outcomes* that are favorable to affected actors.[107] Stewart's analysis of procedural regard is confined to individuals and groups, and explicitly excludes states, which, as Stewart notes, raise different issues.[108] In this book, I address this gap in the literature, and argue that procedural

104 W G Werner, 'The Politics of Expertise: Applying Paradoxes of Scientific Expertise to International Law' in M Ambrus, K Arts, E Hey and H Raulus (eds), *The Role of 'Experts' in International and European Decision-Making Processes: Advisors, Decision Makers or Irrelevant Actors?* (Cambridge University Press, 2014) 44, 56.

105 See generally Stewart, 'Remedying Disregard in Global Regulatory Governance', above n 24.

106 *Ibid* 224–225.

107 *Ibid* 224.

108 *Ibid* 220.

regard also provides a highly apposite lens for states affected by compliance decisions in IEL.

Since the inception of the GAL project, its formative authors have acknowledged that Western, liberal biases may inhere in GAL principles and practices, and the possibility that GAL mechanisms may serve to entrench and perpetuate existing power imbalances.[109] Stewart, for example, recognizes the possibility that GAL procedures will be 'exploited by well-organized economic interests to entrench their influence'.[110] Others take a more definitive, critical stance on this issue. Chimni concludes that GAL has a limited role to play in enhancing the fairness of global governance.[111] This view is informed by TWAIL critiques,[112] which are centrally concerned with the oppressive and transformative potential of the language and discourses of international law from third world[113] perspectives.[114] According to Chimni, GAL represents another manifestation of 'contemporary international law and ... international institutions that have an imperial character'.[115] He argues that 'from a third world perspective, GAL has a limited role in injecting the elements of democracy, equity, and justice into international law and institutions. Indeed, GAL can be co-opted by powerful states to their advantage'.[116]

Thus, while numerous scholars acknowledge the problem associated with disregard for the interests of weaker actors, they provide contrasting views regarding the utility of administrative procedures in addressing this concern. The case-study analysis in this book examines the veracity of these competing perspectives in the settings provided by MEA compliance systems.

109 Kingsbury, Krisch and Stewart, 'The Emergence of Global Administrative Law', above n 1, 51–52.

110 R B Stewart, 'Part I Courts, Institutions, and Access to Justice: Legitimacy and Accountability in Global Regulatory Governance: Global Administrative Law and Developing Countries' (2009) 1 *Jindal Global Law Review* 41, 43–44.

111 Chimni, 'Co-option and Resistance', above n 85, 826; B S Chimni, 'Global Administrative Law: Winners and Losers' (Paper presented at the NYU Law School Global Administrative Law Conference, 22–23 April 2005) 11 <http://iilj.org/global_adlaw/documents/ChimniPaper.pdf>.

112 Chimni, 'Co-option and Resistance', above n 85, 799. There is a diversity of approaches reflected in TWAIL scholarship – for a summary of these approaches, see, eg, O Chinedu Okafor, 'Newness, Imperialism, and International Legal Reform in our Time: A TWAIL Perspective' (2005) 43 *Osgoode Hall Law Journal* 171, 176–177.

113 The third world can be defined as the 'contemporary term for those non-European societies and territories which were colonized from the sixteenth century onwards by the European Empires, and which acquired political independence since the 1940s': A Anghie, *Imperialism, Sovereignty and the Making of International Law* (Cambridge University Press, 2005) 3.

114 J T Gathii, 'TWAIL: A Brief History of its Origins, its Decentralized Network and a Tentative Bibliography' (2011) 3(1) *Trade, Law and Development* 26, 43; K Michelson, 'Critical Approaches' in D Bodansky, J Brunnée and E Hey (eds), *The Oxford Handbook of International Environmental Law* (Oxford University Press, 2007) 262, 273–274.

115 Chimni, 'Co-option and Resistance', above n 85, 800.

116 *Ibid* 826; Chimni, 'Winners and Losers', above n 111, 11.

5.1 Evaluative criteria for procedural regard

The focus of this aspect of the evaluative framework is on if and how administrative procedures promote regard for the compliance interests of affected states. This discussion builds upon the fine-grained criteria for analyzing procedural tools of an administrative character articulated in section 3 above, and extends Stewart's analysis of procedural regard and disregard[117] for marginalized individuals and groups to state actors in international law. The key question is: in what ways do procedures for accountability, transparency, participation, reason giving and review facilitate or obstruct opportunities for affected states' interests to be heard and considered in compliance decision making?

Accountability arrangements may be regard-enhancing if the processes and consequences for non-compliance are tailored in a way that takes into account affected states' individual capacities and circumstances. For example, while a sanction-oriented approach to non-compliance may be deemed suitable for highly industrialized developed states, if the root cause of non-compliance is genuine capacity constraints, a more appropriate focus might be providing financial and technical assistance to facilitate states' return to compliance.[118] This is not to suggest that compliance obligations per se should vary as this would undermine the principle of *pacta sunt servanda* enshrined in Article 26 of the Vienna Convention on the Law of Treaties.[119] Rather, a differentiated approach may apply to compliance *consequences* taking into account country-specific factors.[120] These types of differentiated mechanisms can be contrasted with accountability arrangements that require consistent responses to all non-compliant states, reflecting aims of formal equality, rather than responsiveness to the interests of affected states. Procedural disregard may be evidenced by the imposition of penalties for non-compliance without an opportunity for dialogue in relation to the alleged deficient performance.

Transparency also has the potential to be regard-enhancing, yet financial, technical and language barriers can limit less powerful states' engagement with these opportunities.[121] The availability of high-quality information in accordance

117 Stewart, 'Remedying Disregard in Global Regulatory Governance', above n 24, 224–225.

118 See, for instance, the design of the Kyoto Compliance Committee, discussed in Chapter 5.

119 *Vienna Convention on the Law of Treaties*, opened for signature 23 May 1969, 1155 UNTS 331 (entered into force 27 January 1980). This was confirmed in the 2011 Seabed Mining Advisory Opinion, in which the Seabed Dispute Chamber acknowledged the provisions promoting preferential treatment for developing states, yet noted that a sponsoring state's responsibilities and liabilities 'apply equally to all sponsoring States, whether developing or developed': International Tribunal of the Law of the Sea, *Responsibilities and Obligations of States Sponsoring Persons and Entities with respect to Activities in the Area (Advisory Opinion)*, Seabed Disputes Chamber, Case No 17, 1 Feb 2011, para 158.

120 Borras, 'Comparative Analysis of Selected Compliance Procedures', above n 38, 341 fn 70.

121 In relation to these types of risks pertaining to transparency, see, eg, A Gupta, 'Transparency as Contested Political Terrain: Who Knows What About the Global GMO Trade and Why Does It Matter?' (2010) 10(3) *Global Environmental Politics* 32.

with transparency ideals is a pre-requisite for procedural regard in which the effects of a proposed decision on affected states are taken into account in decision-making processes.[122] However, it is important to be alert to which information is made available internally and externally, and also the extent to which the utility of this information is undermined by states' technical and financial capacity constraints. For example, voluminous information on states' compliance with their IEL commitments may be made available within and outside each regime, yet if interpreting this information requires sophisticated English language skills and technical expertise, the ability of some states to utilize this information in order to meaningfully contribute to compliance determinations will be undermined. While an absence of relevant information signifies procedural disregard, making transparent poor-quality information that is not easily digested by target audiences is also less than ideal, and at best only weakly enhances regard for states' compliance interests.

Avenues for participation can allow state representatives to voice their interests and concerns to decision makers,[123] which aligns with procedural regard by bringing affected interests to the fore. Alternatively, such avenues may be non-existent or dominated by wealthy and powerful states with extensive financial and technical resources at their disposal,[124] entrenching procedural disregard. Thus, the following analysis is attuned to which states are able and enabled to participate, and the nature of their participation.

In addition to states representing their own interests, another way in which affected states' interests can be advocated in compliance decision making is through the participation of civil-society representatives, such as NGOs. However, there is a risk that NGOs' positions on certain issues may be one-sided in favour of Northern priorities,[125] as Raustiala notes:

> Indeed, there is a growing chorus of complaint that sees NGOs as unrepresentative, unaccountable, and often plain uninformed—distorting the democratic process, eroding the authority of elected officials, and excluding the voices of those directly affected by global change in favor of an urban, middle-class minority of armchair radicals, based largely in the industrial world.[126]

Where NGOs and other civil-society representatives participate in MEA compliance processes, it is important to evaluate whether their involvement is skewed toward

122 See the definition of procedural regard in Stewart, 'Remedying Disregard in Global Regulatory Governance', above n 24, 224–225.

123 *Ibid* 235–236, 262.

124 On the risk that participation avenues may be dominated by well-organized and financed interests, see, eg, Chimni, 'Co-option and Resistance', above n 85, 826; Chimni, 'Winners and Losers', above n 111, 11; Harlow, 'The Quest for Principles and Values', above n 21, 211.

125 K Raustiala, 'The "Participatory Revolution" in International Environmental Law' (1997) 21 *Harvard Environmental Law Review* 537, 559; Raustiala, 'NGOs in International Treaty-Making', above n 63, 170–171.

126 Raustiala, 'NGOs in International Treaty-Making', above n 63, 169.

the interests of the industrialized countries of the global North, or whether they give voice to developing countries' perspectives. While both may be regard-enhancing in one sense, the latter is particularly significant from a TWAIL perspective to address the concern that administrative procedures are yet another tool for pursuing hegemonic interests in international law.[127]

The practice of providing reasons for compliance decisions invites decision makers to explain and justify their positions, and has considerable potential to enhance regard for affected states' interests.[128] Generally speaking, articulating the reasons for decisions promotes the similar treatment of analogous cases,[129] which accords with accountability aims of fairness and predictability. Procedural regard requires more than this, namely, acknowledgement of context-specific factors that relate to the cause(s) of non-compliance and/or the appropriate responses to facilitate a return to compliance. Accordingly, the following analysis is alert to implicit or explicit acknowledgement – or its absence – of the interests of affected states in the reasons for compliance decisions. In contrast, a paucity of reasons, or a failure to address the interests of affected states in the reasons provided, evidences procedural disregard.[130]

The availability of avenues of review is also potentially regard-enhancing as it provides an additional opportunity for affected states' interests to be voiced and heard.[131] However, again, capacity constraints may limit less powerful states' ability to take advantage of these opportunities where they exist. If avenues for review are circumscribed, opportunities for procedural regard will be concomitantly limited. The absence of any options for review may signify procedural disregard. In sum, these indicia of procedural regard and disregard furnish criteria for the third element of the proceduralization framework.

Although the primary focus of this book is on procedural inputs into decision making, some consideration of the broader substantive principles that shape decision making is required.[132] Of particular note is the principle of common but differentiated responsibilities (CBDR), which relates to both procedural and substantive regard for developing states' interests in IEL. As previously mentioned, procedural regard is conceptually distinct from substantive regard – while the former relates to the decision-making process, the latter implies outcomes that are advantageous to weaker states. In the context of IEL, issues of substantive regard are explicitly or implicitly intertwined with procedures that are an expression of the CBDR principle.

127 See, eg, Chimni, 'Co-option and Resistance', above n 85, 800.

128 Stewart, 'Remedying Disregard in Global Regulatory Governance', above n 24, 225.

129 See, eg, the precedent-based approach for responding to non-compliance of developing and transition countries in the Montreal compliance system case study in Chapter 4.

130 Stewart, 'Remedying Disregard in Global Regulatory Governance', above n 24, 224.

131 *Ibid* 227.

132 See also J Brunnée, 'International Environmental Law and Community Interests: Procedural Aspects' in E Benvenisti and G Nolte (eds), *Community Obligations in International Law* (forthcoming, 2017) <https://papers.ssrn.com/sol3/papers.cfm?abstract_id=2784701>.

The principle of CBDR has been highly influential in IEL, including in MEA compliance systems. According to this principle, all states have common environmental responsibilities, but due to different social, economic, historical and ecological situations, the manner in which each state meets the responsibilities it assumes should vary accordingly.[133] The principle of CBDR was enshrined in Principle 7 of the Rio Declaration at the first Rio Earth Summit in 1992,[134] and rose to prominence in IEL in the decade between this Summit and the World Summit on Sustainable Development of 2002.[135] The principle of CBDR appears to be losing some of its traction in IEL,[136] however even at the peak of its influence, it did not acquire the legal status of a general principle or customary norm of IEL.[137]

Despite this, the prominence of the CBDR principle in IEL, particularly in the 1990s,[138] helped to shape procedures in MEAs in ways that enhanced regard for developing states' interests.[139] MEAs may differentiate between developed and developing countries with respect to central treaty obligations, implementation and non-compliance responses, and financial and technical assistance,[140] reflecting elements of both substantive and procedural regard for developing states' interests. The CBDR principle has been incorporated into the framework treaties of the Kyoto and Montreal Protocols,[141] and its influence in respect of matters of

133 D Hunter, J Salzman and D Zaelke, *International Environmental Law and Policy* (Foundation Press, 4th ed, 2011) 464.

134 United Nations General Assembly, Rio Declaration on Environment and Development (Annex I), A/CONF.151/26 (vol 1), 14 Jun 1992, reprinted in 31 ILM 874.

135 L Rajamani, 'The Changing Fortunes of Differential Treatment in the Evolution of International Environmental Law' (2012) 88(3) *International Affairs* 605, 606.

136 *Ibid* 614–623. Zahar argues that there has been a shift in the international climate change regime away from a 'CBDR mitigation model' to a 'modified polluter-pays mitigation model': A Zahar, *Climate Change Finance and International Law* (Routledge, 2017) 13.

137 See, eg, T Honkonen, 'The Development of the Principle of Common but Differentiated Responsibilities and its Place in International Environmental Regimes' in T Kuokkanen, E Couzens, T Honkonen and M Lewis (eds), *International Environmental Law-making and Diplomacy: Insights and Overviews* (Routledge, 2016) 160, 179.

138 Rajamani, 'The Changing Fortunes of Differential Treatment', above n 135, 606.

139 See V Kohler, *Le Mécanisme de Contrôle du Respect du Protocole de Kyoto sur les Changements Climatiques: Entre Diplomatie et Droit* (2006) ('describing the obligation to take causes of non compliance fully into account as a fundamental principle of NCPs'), cited by C Pitea, 'The Non-Compliance Procedure of the Aarhus Convention: Between Environmental and Human Rights Control Mechanisms' (2006) 16 *Italian Journal of International Law* 85, 104 fn 107.

140 L Rajamani, *Differential Treatment in International Environmental Law* (Oxford University Press, 2006) 93.

141 See the discussion of the Kyoto Protocol in Chapter 3, section 3. The Montreal Protocol allows a staggered timetable for developing countries to meet their obligations to cease producing and using ozone-depleting substances.

implementation[142] and compliance[143] is evident in numerous MEAs. The role of the CBDR principle in shaping compliance procedures underscores that caution should be applied in seeking to transfer the findings of this study in the field of IEL to other international law contexts where this principle has been less influential.

6 Conclusion

This chapter has proposed a framework that facilitates evaluation of the benefits of adopting administrative procedures in MEA compliance systems. This framework identifies the creation of procedurally constrained spaces for compliance decision making and procedural regard for affected states' interests as desirable aims associated with proceduralization, and specifies concrete criteria for analyzing whether procedural and decisional practices in MEA compliance systems align with these purposes. Specifically, procedural tools from GAL furnish criteria for mapping the degree to which administrative procedures have been incorporated in MEA compliance systems, a spectrum between administrative and political decision-making modes facilitates evaluation of the creation of procedurally constrained zones for decision-making, and indicia of procedural regard and disregard allow assessment of whether and how administrative procedures promote consideration of affected states' compliance interests. In the following chapter, the similarities and differences in the procedural design of the three case-study compliance systems are canvassed, before the proceduralization framework is applied to each system in Chapters 4–6.

142 F Soltau, *Fairness in International Climate Change Law and Policy* (Cambridge University Press, 2009) 184. See, eg, *Convention on Biological Diversity*, opened for signature 5 June 1992, 1760 UNTS 79 (entered into force 29 December 1993), preamble, para 16; *United Nations Convention to Combat Desertification in Those Countries Experiencing Serious Drought and/or Desertification, Particularly Africa*, opened for signature 17 June 1994, 1954 UNTS 3 (entered into force 14 October 1994), arts 3(d), 4(2)(b), 5, 6 and 7; *Agreement for the Implementation of the Provisions of the United Nations Convention on the Law of the Sea of 10 December 1982 Relating to the Conservation and Management of Straddling Fish Stocks and Highly Migratory Fish Stocks*, opened for signature 4 December 1995, 2167 UNTS 3 (entered into force 11 December 2001) preamble, para 8, arts 24(1), 24(2)(c), 26.

143 For example, the compliance procedures adopted under the Biosafety Protocol specify that when deciding upon measures to address non-compliance, the Committee should take into account 'the capacity of the party concerned, especially developing country parties, in particular least developed and small island developing States amongst them, and parties with economies in transition, to comply, and such factors as the cause, type, degree and frequency of non-compliance': J Bulmer, 'Compliance Regimes in Multilateral Environmental Agreements' in J Brunnée, M Doelle and L Rajamani (eds), *Promoting Compliance in an Evolving Climate Regime* (Cambridge University Press, 2012) 55, 61. Similar provisions are also contained in the CITES compliance procedures: see Chapter 6, section 2.4.

3 An introduction to the case study compliance systems

1 Introduction

This chapter focuses on the institutional and procedural apparatus for each of the three MEA compliance-system case studies. This discussion of each compliance system foregrounds four common elements: decision-making bodies, performance-review mechanisms, multilateral non-compliance procedures and non-compliance response measures. The case study selection is justified on the basis of both similarities and differences between these three regimes. The application of matrices developed by Esty demonstrates that their institutional design and the type of issues that they address are particularly likely to raise legitimacy concerns that may be ameliorated by adopting administrative procedures.[1] An overview of the design of each compliance system provides valuable context for the following evaluative process.

2 A brief overview of the Montreal Protocol to the Ozone Convention

The Ozone Convention[2] and its Montreal Protocol[3] are widely regarded as international success stories.[4] The operation of the Montreal compliance system,

1 D C Esty, 'Good Governance at the Supranational Level: Globalizing Administrative Law' (2006) 115 *Yale Law Journal* 1490, 1509–14.

2 *Vienna Convention for the Protection of the Ozone Layer*, opened for signature 22 March 1985, 26 ILM 1529 (entered into force 22 September 1988) ('Ozone Convention').

3 *Montreal Protocol on Substances that Deplete the Ozone Layer 1987*, opened for signature 16 September 1987, 1989 UNTS 28 (entered into force 1 January 1989) ('Montreal Protocol').

4 In 2003, former UN Secretary-General Kofi Annan described the Montreal Protocol as 'perhaps the single most successful international environmental agreement to date': UNEP Ozone Secretariat, 'Key Achievements of the Montreal Protocol to Date' (2012) <http://ozone.unep.org/new_site/en/Information/Information_Kit/Key_achievements_of_the_Montreal_Protocol_2012.pdf>. See also O R Young, 'Effectiveness of International Environmental Regimes: Existing Knowledge, Cutting-Edge Themes, and Research Strategies' (2011) 108(50) *Proceedings of the National Academy of Sciences* 19853, 19853. For a more critical appraisal, see B J Gareau, *From Precaution to Profit: Contemporary Challenges to Environmental Protection in the Montreal Protocol* (Yale University Press, 2013).

reinforced by conditional funding arrangements to support Parties' return to compliance, has significantly contributed to these successes.[5] The history of the Montreal Protocol reflects a gradual increase in, and tightening of, expectations on states over time. Scientific warnings about the accelerating destruction of Earth's stratospheric ozone layer, resulting in dangerous increases in the levels of ultraviolet radiation, catalyzed the United Nations Environment Programme (UNEP) Governing Council to initiate negotiations for a global agreement to protect the ozone layer in 1981.[6] The resultant Vienna Convention for the Protection of the Ozone Layer opened for signature on 22 March 1985, and entered into force on 22 September 1988. Although the obligations in the Ozone Convention are broad and vague, Article 6(4)(h) provides for a more substantial implementing Protocol. The resultant Montreal Protocol, which provides for the progressive elimination of ozone-depleting substances,[7] opened for signature on 16 September 1987, and entered into force on 1 January 1989. Significantly, the Montreal Protocol provides defined timetables for reducing and phasing out, by percentage increments, the consumption and production of specified ozone-depleting substances, which are listed in four annexes.[8] Upon Timor-Leste's ratification on 16 September 2009, the Montreal Protocol became the first global MEA to achieve universal participation with 196 Parties.[9] Currently, both the Ozone Convention and the Montreal Protocol have been ratified by 197 states, making them the first UN treaties to achieve universal ratification.[10]

Through a series of adjustments and amendments, steps have been taken to progressively strengthen the Protocol's requirements. To date, the Protocol's unique adjustment provision under Article 9(2) has been utilized six times, in 1990, 1992, 1995, 1997, 1999 and 2007.[11] Each set of changes has tightened the Protocol's requirements, and has been automatically applicable to all Parties. Unlike adjustments, which do not require ratification by a Party before it is bound by the new requirements, amendments to the Protocol are only applicable to the Parties that have ratified them. In accordance with the procedure specified in Article 9(4), these powers have been utilized five times, in 1990, 1992, 1997,

5 A Cardesa-Salzmann, 'Constitutionalising Secondary Rules in Global Environmental Regimes: Non-Compliance Procedures and the Enforcement of Multilateral Environmental Agreements' (2011) 24(1) *Journal of Environmental Law* 103, 103.

6 U Beyerlin and T Marauhn, *International Environmental Law* (Hart Publishing, 2011) 154.

7 F Romanin Jacur, 'The Non-Compliance Procedure of the 1987 Montreal Protocol to the 1985 Vienna Convention on Substances that Deplete the Ozone Layer' in T Treves et al (eds), *Non-Compliance Procedures and Mechanisms and the Effectiveness of International Environmental Agreements* (TMC Asser Press, 2009) 11, 11.

8 Ozone Convention, above n 2, arts 2 and 3, and annexes.

9 Beyerlin and Marauhn, *International Environmental Law*, above n 6, 155.

10 UNEP Ozone Secretariat, 'The Vienna Convention for the Protection of the Ozone Layer' (2017) <http://ozone.unep.org/en/treaties-and-decisions>.

11 UNEP Ozone Secretariat, 'Adjustments to the Montreal Protocol' (2017) <http://ozone.unep.org/en/handbook-montreal-protocol-substances-deplete-ozone-layer/27594>.

1999 and 2016,[12] to enable, inter alia, the listing of approximately ninety additional substances and their associated reduction timetables, and the creation of a financial mechanism to facilitate developing countries' compliance. Thus, Parties' current obligations under this treaty framework are significantly more onerous than they were at the treaty's inception in the mid-1980s.

The Montreal Protocol is arguably the most effective and successful of the global MEAs,[13] although it has faced some notable challenges. A 2012 document enumerating the Montreal Protocol's achievements in its first twenty-five years states that 'the Protocol has now led to the phase-out of 98 percent of the historic levels of production and consumption of ozone-depleting substances'.[14] Significantly, it is predicted that with ongoing, comprehensive implementation of the Protocol's provisions, the ozone layer will continue to recover and should return to 1980 benchmark levels by mid-century.[15]

Despite these remarkable successes, the Montreal Protocol has faced historical challenges regarding the black market in ozone-depleting substances, particularly in chlorofluorocarbons (CFCs) during the 1990s.[16] A contemporary challenge for the Montreal Protocol is the phasing down of hydrofluorocarbons (HFCs), which are potent greenhouse gases and were widely used as substitutes for hydrochlorofluorocarbons (HCFCs), particularly by developing countries and countries with economies in transition (CEITs), as the latter were phased out under the Protocol. In October 2016, the Parties agreed to the Kigali amendment to the Montreal Protocol that aims to phase down the production and consumption of HFCs, with staged timetables and differentiated baselines for developed and developing countries.[17] This recent development underscores the Protocol's ongoing significance as a regime to protect the ozone layer, and contribute to global efforts to mitigate climate change.[18]

12 While the previous four amendments have been ratified by all state Parties, the Kigali Amendment to the Montreal Protocol was adopted in October 2016, and will enter into force on 1 January 2019 if at least twenty instruments of ratification, acceptance or approval of the Amendment have been deposited by Parties: UNEP, 'Treaties and Decisions' (2017) <http://ozone.unep.org/en/treaties-and-decisions>.

13 See above n 4.

14 UNEP Ozone Secretariat, 'Key Achievements of the Montreal Protocol to Date', above n 4.

15 World Meteorological Organization and United Nations Environment Programme, *Assessment for Decision-Makers: Scientific Assessment of Ozone Depletion: 2014*, World Meteorological Organization Global Ozone Research and Monitoring Project, Report No 56, Geneva, Switzerland <http://ozone.unep.org/Assessment_Panels/SAP/SAP2014_Assessment_for_Decision-Makers.pdf>.

16 E DeSombre, 'The Experience of the Montreal Protocol: Particularly Remarkable, and Remarkably Particular' (2000–2002) 19 *UCLA Journal of Environmental Law and Policy* 49, 62–69.

17 UNEP, Decision XXVIII: Further Amendment to the Montreal Protocol, UNEP/OzL.Pro.28/CRP/10, 14 Oct 2016, and UNEP, Decision XXVIII: Further Amendment to the Montreal Protocol, UNEP/OzL.Pro.28/CRP/11, 14 Oct 2016 (together the 'Kigali amendment').

18 UNEP estimates that phasing down HFCs could prevent up to 0.5 degrees Celsius of global warming by the end of the twenty-first century: UNEP, 'The Kigali Amendment to the

2.1 An introduction to the Montreal compliance system

Like the Montreal Protocol itself, the non-compliance procedure established under Article 8 has been credited as a 'remarkable success', with numerous other MEAs attempting to emulate its effective formula.[19] This sub-section articulates an expansive definition of the Montreal compliance system and outlines the roles played by the key decision-making bodies in this system. Thereafter, a brief overview of the performance-review information, multilateral non-compliance procedures and non-compliance response measures utilized in this compliance system is provided. A distinctive feature of the Montreal compliance system is the extent of reliance on conditional funding to reinforce the procedural mechanisms for non-compliance.

2.1.1 Decision-making bodies

A broad definition of the Montreal compliance system includes but extends beyond the formal non-compliance procedure. The legal basis for the Montreal Protocol's non-compliance procedure is found in Article 8: 'The Parties, at their first meeting, shall consider and approve procedures and institutional mechanisms for determining non-compliance with the provisions of this Protocol and for treatment of Parties found to be in non-compliance.'

The Protocol's non-compliance procedure was adopted by Decision IV/5 of the fourth Meeting of the Parties (MOP) in 1992,[20] and subsequently amended in 1998.[21] The non-compliance procedure consists of only sixteen paragraphs, and thus the practice of the Implementation Committee 'relies to a considerable extent on custom and precedent'.[22] The apparatus that has built up around this non-compliance procedure is significantly more elaborate than these sixteen paragraphs would suggest.

The key bodies involved in implementing the formal non-compliance procedure are the Secretariat, the Implementation Committee and the MOP. On the basis

Montreal Protocol: Another Global Commitment to Stop Climate Change' (2016) <http://web.unep.org/kigali-amendment-montreal-protocol-another-global-commitment-stop-climate-change>. Due to its successes in phasing down ozone-depleting substances, most of which are potent greenhouse gases, the Montreal Protocol has been described as the 'most successful climate treaty to date': M Molina et al, 'Reducing Abrupt Climate Change Risk using the Montreal Protocol and Other Regulatory Actions to Complement Cuts in CO_2 Emissions' (2009) 106(49) *Proceedings of the National Academy of Sciences* 20616, 20617.

19 Cardesa-Salzmann, 'Constitutionalising Secondary Rules in Global Environmental Regimes', above n 5, 103.

20 UNEP, Decision IV/5: Non-Compliance Procedure, UNEP/OzL.Pro.4/1, 25 Nov 1992 ('Non-compliance procedure').

21 UNEP, Decision X/10: Review of the Non-Compliance Procedure, UNEP/OzL.Pro.10/9, 3 Dec 1998.

22 Ozone Secretariat, *Implementation Committee under the Non-Compliance Procedure of the Montreal Protocol on Substances that Deplete the Ozone Layer: Primer for Members* (October 2007) 4 <http://ozone.unep.org/new_site/en/resources.php>.

of performance information received from Parties, the Secretariat prepares an overview of collective trends in Parties' implementation for the MOP, while Party-specific compliance issues are referred to the Implementation Committee. This Committee, which is composed of ten Party representatives elected for two-year terms, meets twice annually. It is an advisory and conciliatory body, the responsibilities of which include managing the Protocol's non-compliance procedure. The Implementation Committee's preliminary decisions on compliance matters are forwarded to the MOP for adoption and approval as interim or final decisions.[23] The MOP, which is a political body composed of representatives of the Parties to the Montreal Protocol, almost always follows the advice of the Implementation Committee,[24] underscoring the significance of this Committee's recommendations for the operation of the compliance system as a whole.

However, the success of the Montreal compliance system is not attributable to the formal reporting system and non-compliance procedure alone – these processes have been buttressed by a broader system of international institutions and procedures.[25] In particular, the Multilateral Fund and the Global Environment Facility have contributed to the compliance system's success by creating financial incentives for developing and transition countries to co-operate with the formal non-compliance procedure. The creation of the Multilateral Fund financial mechanism, which uses voluntary contributions from developed states to defray 'incremental'[26] compliance costs and provide technical assistance for Article 5 developing countries in meeting their obligations under the Protocol, has played a pivotal role in the regime's success.[27] As of May 2017, it was reported that, since 1991, the Multilateral Fund has disbursed approximately US $3.45 billion in support of projects and activities that have contributed to the phasing out of more than 468,000 ODP[28] tonnes of consumption and production of ozone-depleting substances in developing

23 Romanin Jacur, 'The Non-Compliance Procedure of the 1987 Montreal Protocol', above n 7, 26.

24 Ozone Secretariat, *Primer for Members,* above n 22, 18.

25 O Greene, 'The System for Implementation Review in the Ozone System' in D G Victor, K Raustiala and E B Skolnikoff (eds), *The Implementation and Effectiveness of International Environmental Commitments* (MIT Press, 1998) 89; G de Burca, R O Keohane and C F Sabel, 'Global Experimentalist Governance' (2014) 44(3) *British Journal of Political Science* 477.

26 The 'incremental costs' are 'the extra costs between the costs of the alternative case and the baseline case': G M Bankobeza, *Ozone Protection: The International Legal Regime* (Eleven International Publishing, 2005) 228 at fn 39; see also UNEP, Indicative List of Categories of Incremental Costs, Report of the Fourth Meeting of the Parties to the Montreal Protocol, UNEP/OzL.Pro.5/15, 25 Nov 1992, Annex VIII.

27 K Raustiala, *Reporting and Review Institutions in 10 Multilateral Environmental Agreements* (UNEP, 2001) 33 <www.peacepalacelibrary.nl/ebooks/files/C08-0025-Raustiala-Reporting.pdf>.

28 Ozone-depleting potential (ODP) is a measure for the amount of ozone depletion caused by a chemical substance, calculated as a ratio of the impact on ozone of the substance compared to the impact of a similar mass of CFC-11.

countries.[29] Similarly, the Global Environment Facility, which was established to support action on four global environmental issues including ozone layer depletion, utilizes a 'Trust Fund' administered by the World Bank and provides financial assistance to CEITs. These countries are ineligible for funding support from the Multilateral Fund as they are not developing countries,[30] yet may still face difficulties in meeting developed countries' phase-out schedules under the Protocol.[31]

Accordingly, the analysis of the Montreal compliance system in this book includes the roles of the Secretariat, the Implementation Committee, the MOP, the Multilateral Fund and the Global Environment Facility as 'subsidiary bodies of an administrative character'[32] that comprise the Montreal compliance system. As is shown in Chapter 4, the roles played by the Multilateral Fund and Global Environment Facility in providing conditional financial assistance to support non-compliant Parties' return to compliance have been pivotal in buttressing the procedural regard opportunities available in the compliance system.

2.1.2 Performance-review mechanisms

The Secretariat is the focal point for the collection and processing of performance-review information in the Montreal compliance system. National reports are the primary source from which the Secretariat gathers, collates and synthesizes Parties' performance-review data. On an annual basis, Parties to the Protocol are required to provide statistical data pertaining to the production, consumption, import and export of controlled ozone-depleting substances relative to the baseline year and the year since the Party ratified the Protocol, including under the Protocol's essential-use exemption.[33] Thus, there is significant reliance placed upon state-supplied compliance data.

It is common practice for the Secretariat to request clarification from a Party in situations in which anomalous or inconsistent data is identified.[34] Furthermore,

29 Multilateral Fund for the Implementation of the Montreal Protocol, 'Welcome to the Multilateral Fund for the Implementation of the Montreal Protocol' (2017) <www.multilateralfund.org/default.aspx>.

30 As Article 5 of the Montreal Protocol defines eligibility in terms of per capita consumption of ozone-depleting substances, a small number of CEITs are eligible for Multilateral Fund assistance: Bankobeza, *Ozone Protection*, above n 26, 235.

31 O Yoshida, 'Soft Enforcement of Treaties: The Montreal Protocol's Noncompliance Procedure and the Functions of Internal International Institutions' (1999) 10(1) *Colorado Journal of International Environmental Law and Policy* 95, 100.

32 B Kingsbury, N Krisch and R B Stewart, 'The Emergence of Global Administrative Law' (2005) 68 *Law and Contemporary Problems* 15, 21.

33 Montreal Protocol, above n 3, art 7; Raustiala, 'Reporting and Review Institutions', above n 27, 34.

34 Such anomalies and inconsistencies can sometimes be identified as a result of additional information collected through informal implementation review activities undertaken by a range of related bodies, including the Executive Committee of the Multilateral Fund, the

the Global Ozone Observing System,[35] which was established by the World Meteorological Organization (WMO) and contributes information on the atmospheric concentrations of ozone through the use of approximately 140 monitoring stations worldwide, provides a 'very rough and indirect check' of Parties' reports.[36] There are no other formal review or verification measures used in relation to Parties' reported information.

2.1.3 Multilateral non-compliance procedures

There are multiple ways in which the Montreal Protocol's non-compliance procedure can be triggered. The causes of non-compliance can be either procedural, such as a failure to comply with national reporting requirements,[37] and/or relate to substantive obligations regarding phasing down the production and consumption of ozone-depleting substances,[38] and controlling international trade in such substances.[39] Where compliance concerns exist, the non-compliance procedure can be invoked: (1) by the Secretariat based on a Party's annual reports,[40] (2) by a Party concerned about another Party's potential non-compliance[41] and/or (3) by the self-reporting of a Party which, 'despite having made its best, bona fide efforts', is, or anticipates it will be, in non-compliance.[42] Overwhelmingly, the most common trigger in practice is (1). The trigger in (3) was used in the mid-1990s by the Russian Federation and other CEITs,[43] but otherwise has been utilized infrequently.

If the Secretariat becomes aware of non-compliance issues that it cannot resolve through 'administrative action or diplomatic contacts', the matter is forwarded for consideration by the Implementation Committee.[44] Once a Party is referred to the Implementation Committee through one of these three methods, it can be

Technology and Economic Assistance Panel, and the Global Environment Facility: Raustiala, 'Reporting and Review Institutions', above n 27, 34–35.

35 Global Ozone Observing System, 'Global Observing System Information Centre' (2017) <www.ncdc.noaa.gov/gosic>.

36 J Wettestad, 'Enhancing Climate Compliance – What are the Lessons to Learn from Environmental Regimes and the EU?' in O Schram Stokke, J Hovi and G Ulfstein (eds), *Implementing the Climate Regime: International Compliance* (Earthscan, 2005) 209, 214–215.

37 Montreal Protocol, above n 3, arts 7 and 9.

38 *Ibid* arts 2 and 3 and annexes.

39 *Ibid* art 4; Yoshida, 'Soft Enforcement of Treaties', above n 31, 97, 127.

40 Non-compliance procedure, above n 20, para 3.

41 *Ibid* para 1.

42 *Ibid* para 4.

43 See further Chapter 4, section 2. As Tallberg notes, due to the technical and state-controlled nature of the information on the control and phasing out of ozone-depleting substances, it is unlikely that the Party-to-Party trigger in (2) will be utilized: J Tallberg, 'Orchestrating Enforcement: International Organizations Mobilizing Compliance Constituencies' in K W Abbott, P Genschel, D Snidal and B Zangl (eds), *International Organizations as Orchestrators* (Cambridge University Press, 2015) 166, 179–180.

44 Ozone Secretariat, *Primer for Members,* above n 22, 10.

requested to attend the next Committee meeting to explain its situation. Parties' explanations to the Committee underpin a 'non-confrontational but frank' discussion characterized by pragmatism and flexibility, rather than legal argumentation.[45] It is at this stage of the non-compliance process that the Multilateral Fund and Global Environment Facility may be invited to advise both the Implementation Committee and the Party concerned on the financial and technical assistance available to support the Party's return to compliance.

2.1.4 Non-compliance response measures

The non-compliance procedure provides for escalating response measures that the MOP can utilize, typically upon the advice of the Implementation Committee. Decision IV/5 contains the following Indicative List of Measures that may be taken in relation to Parties' non-compliance:

a) appropriate assistance, including assistance for the collection and reporting of data, technical assistance, technology transfer and financial assistance, information transfer and training;
b) issuing cautions;
c) suspension, in accordance with the applicable rules of international law concerning suspension of the operation of the treaty, of specific rights and privileges under the Protocol, whether or not subject to time limits, including those concerned with industrial rationalization, production, consumption, trade, transfer of technology and institutional arrangements.[46]

This may be seen as a type of 'flexible pyramid', with responses reflecting a graduated scale of increasingly strong measures.[47] The Ozone Secretariat reports that options (a) and (b) have been utilized by the MOP, acting on the Committee's recommendations, and that option (c) has 'not been implemented to date, but has been referred to in a cautionary context'.[48] At the seventh MOP (Vienna, 1995), restrictions on trade in ozone-depleting substances were imposed upon

45 Raustiala, 'Reporting and Review Institutions', above n 27, 36; J Klabbers, 'Compliance Procedures' in D Bodansky, J Brunnée and E Hey (eds), *The Oxford Handbook of International Environmental Law* (Oxford University Press, 2007) 995, 996–997.

46 UNEP, Decision IV/5, above n 20, Annex V.

47 B Kingsbury, 'The Concept of Compliance as a Function of Competing Conceptions of International Law' (1998) 19 *Michigan Journal of International Law* 345, 364, citing I Ayres and J Braithwaite, *Responsive Regulation: Transcending the Deregulation Debate* (Oxford University Press, 1992) 35–40.

48 Ozone Secretariat, *Primer for Members*, above n 22, 45. See also Ozone Secretariat, *Handbook for the Montreal Protocol on Substances that Deplete the Ozone Layer, Tenth Edition* (2016) <http://ozone.unep.org/sites/ozone/files/Publications/Handbooks/MP-Handbook-2016-English.pdf>.

Belarus, Ukraine and the Russian Federation,[49] which appear to go beyond the three listed non-compliance responses. Fitzmaurice argues that this was in fact 'the imposition of a new restriction on the countries' right to trade with other Parties which they enjoyed quite independently of the Protocol, not the suspension of a right conferred by the Protocol'.[50] Arguably, therefore, the range of non-compliance responses relied upon in practice exceeds the options in the Indicative List of Measures. Despite this, the general tenor of non-compliance proceedings under the Montreal Protocol has been facilitative, flexible and focused on understanding the reasons for states' compliance issues.

3 A brief overview of the UNFCCC and the Kyoto Protocol

The Kyoto compliance system has been described as 'the most ambitious and elaborate of the MEA compliance regimes in operation today'.[51] The Kyoto Protocol[52] is a treaty under the aegis of the UNFCCC[53] that has achieved almost universal membership.[54] The UNFCCC enshrines differential treatment for Annex I[55] industrialized countries and non-Annex I developing countries. All Parties have procedural obligations to publish national emissions inventories, and to formulate, publish and regularly update national and regional programmes containing

49 UNEP, Decision VII/17: Compliance with the Montreal Protocol by Belarus, UNEP/OzL.Pro.7/12, 27 Dec 1995; UNEP, Decision VII/18: Compliance with the Montreal Protocol by Russia, UNEP/OzL.Pro.7/12, 27 Dec 1995; UNEP, Decision VII/19: Compliance with the Montreal Protocol by Ukraine, UNEP/OzL.Pro.7/12, 27 Dec 1995.

50 M Fitzmaurice, 'Non-Compliance Procedures and the Law of Treaties' in T Treves et al (eds), *Non-Compliance Procedures and Mechanisms and the Effectiveness of International Environmental Agreements* (Cambridge University Press, 2009) 453, 472–473. Cf Werksman's explanation that the Montreal Protocol created a general ban on trade in regulated substances between Parties and non-Parties, which 'allows the NCP to treat the ability to trade in regulated substances as a privilege granted by the Montreal Protocol and to incrementally suspend that privilege in order to encourage compliance': J Werksman, 'Compliance and the Kyoto Protocol: Building a Backbone in a Flexible Regime' (1998) 9(1) *Yearbook of International Environmental Law* 48, 72.

51 J Brunnée, 'Climate Change and Compliance and Enforcement Processes' in R Rayfuse and S V Scott (eds), *International Law in the Era of Climate Change* (Edward Elgar, 2012) 290, 303.

52 *The United Nations Framework Convention on Climate Change*, opened for signature 9 May 1992, 1771 UNTS 107 (entered into force 21 March 1994) ('UNFCCC').

53 *Kyoto Protocol to the United Nations Framework Convention on Climate Change*, opened for signature 11 December 1997, 37 ILM 22 (entered into force 16 February 2005) ('Kyoto Protocol').

54 The UNFCCC has 197 Parties in 2017: UNFCCC, 'About UNFCCC' (2017) <http://newsroom.unfccc.int/about/>.

55 That is, developed countries and countries in transition. Other categories of countries include Annex II countries, which include members of the Organization for Economic Co-operation and Development and the European Community, and non-Annex I Parties, which are developing countries.

mitigation measures.[56] It is only Annex I industrialized countries that are required to adopt national mitigation policies for limiting greenhouse gas (GHG) emissions and protecting and enhancing sinks and reservoirs with the aim of reducing overall GHG emissions to 1990 levels.[57]

The original Framework Convention on Climate Change was soon buttressed by a more detailed protocol. Two years after the entry into force of the UNFCCC, the Kyoto Protocol to the UNFCCC was adopted at the third COP in 1997. The Kyoto Protocol mirrors the UNFCCC's asymmetric approach for Annex I and non-Annex I Parties. The Kyoto Protocol commits each Annex I Party to achieve differentiated, legally binding emissions targets of at least 5 percent below 1990 levels in the first commitment period (2008–2012). In addition to reducing their GHG emissions, or increasing their sinks and reservoirs, Parties are able to meet their emissions targets through the Kyoto Protocol's innovative flexibility mechanisms – namely, joint implementation, the clean development mechanism and emissions trading.[58] The aim of these flexibility mechanisms is to allow reductions to be achieved where it is most efficient to do so from an economic perspective, reducing the overall costs of emissions reductions.[59] In contrast, non-Annex I developing countries are only required to meet certain procedural obligations, such as reporting. The developed countries of the Organization for Economic Co-operation and Development and the European Community listed in Annex II are further required to transfer financial resources and technology to developing countries to assist them with fulfilling their procedural duties.[60] The UNFCCC and Kyoto Protocol's differentiated approach to central treaty obligations, which apply to some countries and not others, is unique amongst global MEAs.[61]

Despite this codification of differential treatment, the meaning and appropriate implementation of the regime's core principle of common but differentiated responsibilities (CBDR) have remained deeply contested. The North–South tensions that constitute a fault line in international climate politics occur against a complex background characterized by the industrialized North's primary responsibility for historical emissions, national variations in wealth and emissions use and the disproportionate vulnerability of the poorest populations within developing countries to adverse climate-change impacts.[62] Indeed, the Kyoto Protocol has been riven by North–South divides that partially account for its hampered effectiveness to date, and strongly influenced its replacement by the Paris Agreement in 2015. Concerns about differential treatment informed the

56 UNFCCC, above n 52, art 4(1)(a), (b).
57 *Ibid* art 4(2)(a), (b).
58 *Ibid* arts 6, 12 and 17, respectively.
59 Beyerlin and Marauhn, *International Environmental Law*, above n 6, 162.
60 UNFCCC, above n 52, art 4(3).
61 L Rajamani, J Brunnée and M Doelle, 'Introduction: The Role of Compliance in an Evolving Climate Regime' in J Brunnée, M Doelle and L Rajamani (eds), *Promoting Compliance in an Evolving Climate Regime* (Cambridge University Press, 2012) 1, 3.
62 *Ibid* 1–2.

United States' refusal to ratify the Kyoto Protocol in 2001,[63] and the Protocol's emission-reduction commitments do not apply to non-Annex I countries, such as China, India and Brazil, which rank among the world's top ten contributors to cumulative global emissions.[64] For developing and developed countries alike, achieving deep emissions cuts poses financial, regulatory and technical capacity challenges. The lack of a level playing field among major economies' commitments to emissions reductions proved to be a significant obstacle to the Kyoto Protocol's longevity.

The Kyoto Protocol's second commitment period will run from 2013 to 2020. At the eighth Conference of the Parties serving as the Meeting of the Parties to the Kyoto Protocol (CMP) in Doha in December 2012, a reduced number of states, accounting for 22 per cent of global emissions,[65] committed to the second commitment period. This allows for a continuation of the Protocol's legal requirements and preserves the flexibility, accounting, review and compliance mechanisms established under the first commitment period. However, the Doha Amendment is yet to legally enter into force.[66] In the light of the adoption of the Paris Agreement, it is unlikely that the Kyoto Protocol will continue its operation past the end of the second commitment period in 2020. Nonetheless, analysis of its operation to date provides valuable insights regarding the benefits and limits of proceduralization in global environmental governance, which can inform procedural design for the Paris Agreement's compliance apparatus.[67]

3.1 An introduction to the Kyoto compliance system

3.1.1 Decision-making bodies

In the Kyoto compliance system, decision-making power is concentrated in the expert review teams (ERTs) and the Compliance Committee, consisting of the Facilitative and Enforcement Branches. ERTs are small teams of technical experts who review states' national emissions reports and play a pivotal role in triggering non-compliance proceedings in the Compliance Committee. This

63 *Ibid* 3, citing the text of a letter from the President to Senators Hagel, Helms, Craig and Roberts, the White House, Office of the Press Secretary (13 March 2001).

64 *Ibid* 4.

65 This figure was calculated by Rajamani: L Rajamani, 'The Durban Platform for Enhanced Action and the Future of the Climate Regime' (2012) 61(2) *International and Comparative Law Quarterly* 501, 516.

66 United Nations Treaty Collection, 'Chapter XXVII Environment: 7.c Doha Amendment to the Kyoto Protocol' (2017) <https://treaties.un.org/pages/ViewDetails.aspx?src=TREATY&mtdsg_no=XXVII-7-c&chapter=27&clang=_en>. In order for this amendment to enter into force, an instrument of acceptance must be received by the Depositary from at least three-fourths of the Parties to the Protocol: Kyoto Protocol, above n 53, art 20(4).

67 See further Chapter 7, section 3.

Committee, which is also comprised of experts, was established as the body responsible for resolving compliance issues and determining the consequences of non-compliance under Decision 27/CMP.1.[68] Within the Committee, the roles of the Facilitative and Enforcement Branches are bifurcated to reflect the compliance procedures and mechanisms' various facilitation, compliance promotion and enforcement-oriented aims.[69] The Facilitative Branch is tasked with advising on, and facilitating implementation for, all Parties, and promoting Annex I Parties' compliance with Protocol commitments that do not relate to emission-reduction commitments,[70] taking into account the principle of 'common but differentiated responsibilities and respective capacities'.[71] In contrast, the Enforcement Branch has a mandate to take significantly stronger measures in response to questions involving emission-reduction commitments and related reporting and eligibility requirements, 'taking into account the cause, type, degree and frequency of the non-compliance of that Party'.[72] The plenary serves as a link between the Compliance Committee and the CMP, and plays a largely administrative role.[73] The Kyoto compliance system's reliance on technical experts is a key differentiating factor from its counterparts under the Montreal Protocol and CITES.

3.1.2 Performance-review mechanisms

Compared to other MEAs,[74] the measurement, reporting and verification provisions in Articles 5, 7 and 8 of the Kyoto Protocol represent a sophisticated approach to collecting performance-review information. Under Article 5, Parties are required to establish national systems to estimate GHG inventories and removals using the common metric of carbon dioxide equivalents. Tiered methodological approaches to preparing emissions inventories are provided by specified guidance materials produced by the Intergovernmental Panel on Climate Change, the use of which was mandatory in the first commitment period.[75] Article 7

68 Decision 27/CMP.1, Procedures and Mechanisms Relating to Compliance under the Kyoto Protocol, FCCC/KP/CMP/2005/8/Add.3, 30 Mar 2006, section II.

69 Decision 27/CMP.1 states that the procedures and mechanisms' objectives are to 'facilitate, promote and enforce compliance': *ibid* section I.

70 Stated differently, the mandate of the Facilitative Branch is to address questions of implementation that are not within the purview of the Enforcement Branch: G Ulfstein and J Werksman, 'The Kyoto Compliance System: Towards Hard Enforcement' in O S Stokke, J Hovi and G Ulfstein (eds), *Implementing the Climate Regime: International Compliance* (Earthscan, 2005) 39, 45.

71 Decision 27/CMP.1, above n 68, section IV, paras 4 and 6.

72 *Ibid* section XV, para 1.

73 R Lefeber and S Oberthür, 'Key Features of the Kyoto Protocol's Compliance System' in J Brunnée, M Doelle and L Rajamani (eds), *Promoting Compliance in an Evolving Climate Regime* (Cambridge University Press, 2012) 77, 81–2.

74 See, eg, Raustiala, 'Reporting and Review Institutions', above n 27. See also Chapters 4 and 6.

75 Kyoto Protocol, above n 53, art 5(2); A Herold, 'Experiences with Articles 5, 7, and 8: Defining the Monitoring, Reporting and Verification System under the Kyoto Protocol'

stipulates requirements for Parties' submission of national emissions information, the review of which is then conducted by independent third-party ERTs under Article 8. In practice, a primary focus of such review processes is the completeness and reliability of national emissions inventories,[76] which are arguably the 'foundation on which the rest of the international climate regime is built'.[77]

3.1.3 Multilateral non-compliance procedures

The Kyoto compliance system has a formal non-compliance procedure, which is most commonly triggered by ERT reports. The legal basis for the non-compliance procedure is Article 18 of the Kyoto Protocol, which mandates the development of 'appropriate and effective procedures and mechanisms to determine and address cases of non-compliance', including through 'the development of an indicative list of consequences'. After complicated and occasionally fraught negotiations, the Protocol's non-compliance procedure was adopted by Decision 27/CMP.1[78] at the first CMP in 2006. Several key procedural elements were subsequently clarified in the Rules of Procedure of the Compliance Committee to the Kyoto Protocol adopted by Decision 4/CMP.2[79] at the second CMP in 2007.

Decision 27/CMP.1 specifies three ways in which the non-compliance procedure can be triggered: (1) by an ERT report, (2) by the self-nomination of a Party who is not in compliance and (3) by one Party with respect to another Party if the initiating Party provides 'corroborating information'.[80] In practice, the role of ERTs in initiating non-compliance proceedings in the Compliance Committee has proved to be vital.[81] Within seven days of the non-compliance procedure being triggered, the Bureau allocates the matter to the appropriate branch of the Compliance Committee, which conducts a hearing to inform a decision on the relevant compliance issues.

3.1.4 Non-compliance response measures

In the event of a finding of non-compliance, the Facilitative and Enforcement Branches have recourse to a range of tiered responses. The Facilitative Branch may

in J Brunnée, M Doelle and L Rajamani (eds), *Promoting Compliance in an Evolving Climate Regime* (Cambridge University Press, 2012) 122, 125.

76 A Zahar, 'Does Self-Interest Skew State Reporting of Greenhouse Gas Emissions? A Preliminary Analysis Based on the First Verified Emissions Estimates Under the Kyoto Protocol' (2010) 1 *Climate Law* 313, 315.

77 A Zahar, J Peel and L Godden, *Australian Climate Law in Global Context* (Cambridge University Press, 2012) 96.

78 Decision 27/CMP.1, above n 68.

79 Decision 4/CMP.2, Compliance Committee, FCCC/KP/CMP/2006/10/Add.1, 4 Mar 2007.

80 Decision 27/CMP.1, above n 68, section XI, para 1.

81 Lefeber and Oberthür, 'Key Features of the Kyoto Protocol's Compliance System', above n 73, 86.

recommend a number of 'soft' responses, including provision of advice regarding implementation, financial and technical assistance and the formulation of recommendations.[82] The significantly more intrusive consequences of a finding of non-compliance available to the Enforcement Branch include the requirement of a 'compliance action plan' for remedying non-compliance with methodological and reporting requirements,[83] states' suspension from participating in the Protocol's flexibility mechanisms if the non-compliance issue concerns the eligibility requirements[84] and deductions from future emissions allocations if a Party's emission target is exceeded.[85] A Party can appeal to the CMP against a decision of the Enforcement Branch if it believes it has been denied due process and the decision 'relates to' Article 3(1) of the Kyoto Protocol regarding national emissions targets.[86] The Kyoto compliance system is thus a sophisticated administrative apparatus, which privileges the role of expert inputs, for holding states to account for their international environment commitments under the Kyoto Protocol.

4 A brief overview of CITES

The Convention on International Trade in Endangered Species of Wild Fauna and Flora,[87] widely referred to as CITES, entered into force in 1975. As such, it is an older style MEA compared to both the Montreal and Kyoto Protocols. A key aim of CITES is to protect endangered fauna and flora from over-exploitation by regulating commercial wildlife trade across international borders.[88] It is estimated that the annual value of illicit trade in endangered species is worth US $20 billion per year, making it the second largest international black market behind illicit drug smuggling.[89] CITES is not a priori anti-trade, but rather aims to regulate trade in endangered species by requiring the authorization of 'all import, export, re-export and introduction' in specimens of listed species through a licensing system.[90] These specimens are listed on one of three appendices, which together cover approximately 5,600 species of animals and 30,000 species of plants.[91]

82 Decision 27/CMP.1, above n 68, section XIV.
83 *Ibid* section XV, para 5(b).
84 *Ibid* section XV, para 5(c).
85 Specifically, a 'deduction from the Party's assigned amount for the second commitment period of a number of tonnes equal to 1.3 times the amount in tonnes of excess emissions': *ibid* section XV, para 5(a).
86 *Ibid* section XI, para 1.
87 *Convention on International Trade in Endangered Species of Wild Fauna and Flora 1973*, opened for signature 3 March 1973, 993 UNTS 243 (entered into force 1 July 1975) ('CITES').
88 *Ibid* art 2.
89 N South and T Wyatt, 'Comparing Illicit Trades in Wildlife and Drugs: An Exploratory Study' (2011) 32(6) *Deviant Behavior* 538, 540.
90 CITES, above n 87, arts 3, 4 and 5.
91 CITES, 'The CITES Species' (2017) <http://www.cites.org/eng/disc/species.php>.

The tripartite CITES system created by Article II of the Convention envisages varying degrees of protection depending on the rarity of different types of species. Appendix I lists specimens[92] of species[93] that are threatened with extinction and affected by trade; for these specimens 'particularly strict regulation' must apply, and trade 'must only be authorized in exceptional circumstances'[94] where export will not be 'detrimental' to the relevant species' survival.[95] The stringency of these and related requirements in Article III(2) and (3) effectively 'shut[s] down international trade' in Appendix I species.[96] Similar export, but less rigorous import, requirements apply for specimens of species listed in Appendix II,[97] which could be threatened with extinction if their trade is not subject to 'strict regulation' designed 'to avoid utilization incompatible with their survival'.[98] Appendix III lists specimens of species facing over-exploitation in a particular country.[99] The import and export safeguards required in relation to these species are significantly diluted compared to their counterparts in Appendices I and II.[100] Once a species is listed, it is incumbent upon the Parties to implement domestic laws to prohibit trade in specimens in violation of the Convention, including mechanisms for enforcement, confiscation and return.[101]

4.1 An introduction to the CITES compliance system

4.1.1 Decision-making bodies

Compared to its counterparts under the Montreal and Kyoto Protocols, the procedural apparatus under the CITES compliance system has developed in an ad hoc manner over time, and is more fragmented. The responsibility for administering the CITES compliance procedures, which were introduced in 2007, is chiefly shared among three bodies: the Secretariat, the Standing Committee and the COP.[102] The role of the Secretariat, which is administered by UNEP, is pivotal.

92 Specimens are defined as 'animal[s] or plant[s], whether dead or alive': CITES, above n 87, art 1(b).
93 Species are defined to mean 'any species, subspecies or geographically separate population thereof': *ibid* art 1(a).
94 *Ibid* art II(1).
95 *Ibid* art III(2)(a)) and III(3)(a).
96 E Brown Weiss, 'The Five International Treaties: A Living History' in E Brown Weiss and H K Jacobson (eds), *Engaging Countries: Strengthening Compliance with International Environmental Accords* (MIT Press, 1998) 89, 107.
97 CITES, above n 87, art IV.
98 *Ibid* art II(2)(a).
99 *Ibid* art II(3).
100 *Ibid* art V(2) and (3); Brown Weiss, 'The Five International Treaties', above n 96, 106; Beyerlin and Marauhn, *International Environmental Law*, above n 6, 185.
101 CITES, above n 87, art VIII(1).
102 The compliance responsibilities of the Secretariat, Animals and Plants Committees, Standing Committee and COP are specified in paragraphs 9–14 of the CITES compliance

The relevant provisions pertaining to the Secretariat's compliance roles are contained in Article 12(2) of the Convention, and include the power to study Parties' reports and request further information, invite Parties' attention to matters including compliance concerns and the ability to make recommendations for the implementation of the Convention.[103] In addition, the CITES compliance procedures specify that the Secretariat's role in relation to compliance includes, inter alia, advising and assisting Parties on compliance issues, making recommendations for achieving compliance and monitoring the implementation of compliance-related decisions.[104] Reports prepared by the Secretariat for Standing Committee meetings have been primarily responsible for 'triggering' most non-compliance matters.[105] The vesting of extensive compliance powers, including recommendatory functions, in the Secretariat is unusual among the global MEAs, where the Secretariat's primary role in relation to compliance is typically to facilitate information exchange.

While the Secretariat is tasked with identifying instances of non-compliance, the responsibility for handling 'general and specific compliance matters' and 'taking compliance measures' falls to the Standing Committee.[106] The Standing Committee is a political body whose powers have increased extensively since it was established by permanent COP resolution in 1979.[107] It is composed of nineteen state representatives from six geographical regions and was created to oversee the efficient functioning of the Convention – the ambit of which includes but extends well beyond compliance – between COP meetings. In most instances, the Standing Committee follows the Secretariat's recommendations for non-compliance measures, underscoring the Secretariat's significant influence.[108]

procedures, supplementing the original guidance provided in the Convention: CITES, CITES Compliance Procedures, Resolution Conf 14.3, Jun 2007, Annex, para 14(a)–(e) ('CITES compliance procedures'). The compliance functions of the Animals and Plants Committees pertain to recommending remedial action when certain species are subject to unsustainable trade. The work of these two bodies is not central to the operation of the compliance system analyzed in Chapter 6, and thus will not be a focus here.

103 CITES, above n 87, art 12, para 2(d), (e) and (h).

104 CITES compliance procedures, above n 102, Annex, para 14(a)–(e).

105 One exception to this was the case of Bolivia in 1985, which was triggered by a resolution sponsored by Latin American countries whose wildlife populations were adversely affected by Bolivia's illicit trade: R Reeve, 'The Convention on International Trade in Endangered Species of Wild Fauna and Flora' in G Ulfstein, T Marauhn and A Zimmermann (eds), *Making Treaties Work: Human Rights, Environment and Arms Control* (Cambridge University Press, 2007) 134, 140.

106 CITES compliance procedures, above n 102, para 12.

107 CITES, Establishment of the Standing Committee of the Conference of the Parties, Resolution Conf 2.2, 1979 (Rev COP 15).

108 R Reeve, *Policing International Trade in Endangered Species: The CITES Treaty and Compliance* (Earthscan, 2002) 258; R Sandford, 'International Environmental Secretariats: A Case of Neglected Potential?' (1996) 16(1) *Environmental Impact Assessment Review* 3, 8.

The COP, which is composed of government representatives from 183 member countries, is the Convention's main decision-making body,[109] yet its role in relation to compliance is somewhat circumscribed. The COP's role in the CITES compliance system includes providing general policy guidance on compliance issues, directing and overseeing the handling of compliance matters and reviewing Standing Committee decisions on compliance matters 'as needed'.[110] In addition, the COP's decisions to list species on Appendix I or II shape the parameters of non-compliance by either prohibiting or restricting trade in certain species. On one occasion in 1985 the COP played a proactive role in recommending trade sanctions in response to serious generalized non-compliance in Bolivia,[111] but generally the Standing Committee is primarily responsible for decisions regarding non-compliance and the consequences thereof.

A distinctive feature of CITES vis-à-vis the Montreal and Kyoto Protocols is the significant participation rights granted to civil-society representatives, including NGOs.[112] When the CITES Convention opened for signature in 1973, it was atypical in terms of its privileging of NGO participation at COP meetings.[113] In recent times, there has been extensive non-decisional participation by NGOs in the Secretariat,[114] Standing Committee[115] and COP. In addition, NGOs serve as intermediaries engaged to collect performance-review information.[116] As is explored in Chapter 6, NGOs' roles also include championing various listing decisions that alter the parameters for non-compliance under the regime. In some instances, this involvement enhances interest representation for states affected by compliance decisions, and thus procedural regard.

109 C Fuchs, 'Convention on International Trade in Endangered Species of Wild Fauna and Flora (CITES) – Conservation Efforts Undermine the Legality Principle' (2008) 9(11) *German Law Journal* 1565, 1571.

110 CITES compliance procedures, above n 102, para 10.

111 Reeve, 'The Convention on International Trade in Endangered Species of Wild Fauna and Flora', above n 105, 141.

112 S Charnovitz, 'Nongovernmental Organizations and International Law' (2006) 100 *American Journal of International Law* 348, 369–370.

113 K Raustiala, 'The "Participatory Revolution" in International Environmental Law' (1997) 21 *Harvard Environmental Law Review* 537, 547.

114 Article XII(1) of the Convention provides that the Secretariat may be assisted by 'suitable' national, inter-governmental or non-governmental bodies that are 'technically qualified in protection, conservation and management of wild flora and fauna'.

115 Prior to 2003, meetings of the Standing Committee were closed to all except contracted NGOs – the International Union for the Conservation of Nature (IUCN), the United Nations Environment Programme World Conservation Monitoring Centre (UNEP-WCMC) and the Trade Records Analysis of Fauna and Flora in Commerce (TRAFFIC). Since this time, other NGOs with expertise in wildlife issues have been permitted to apply to attend and participate in Standing Committee meetings: R Reeve, 'Wildlife Trade, Sanctions and Compliance: Lessons from the CITES Regime' (2006) 82(5) *International Affairs* 881, 883.

116 Tallberg, 'Orchestrating Enforcement', above n 43, 178–180. In particular, TRAFFIC has played a pivotal role in monitoring wildlife trade and raising awareness of potential compliance issues.

4.1.2 Performance-review mechanisms

A distinctive feature of the CITES compliance system is the Secretariat's recourse to a wide range of sources in compiling review information. The CITES compliance procedures list annual and biennial reports of trade in CITES-listed species, legislative texts as well as other special reports and responses to information requests as 'primary, but not exclusive' sources of performance-review information.[117] Parties' self-reporting is complemented by additional measures to identify infractions, including information from CITES trade databases and registers, NGOs and other public sources such as Internet sites and the media.[118] Interpol and the World Customs Organization play roles in investigating illicit trade and trafficking rings involving Parties and private actors.[119] This range of sources and wealth of information that the Secretariat can consult is significantly greater than in other compliance systems, such as the Montreal and Kyoto Protocols, which frequently rely on state-supplied data. Therefore, despite gaps in the information available, the Secretariat is empowered to meaningfully review the accuracy and completeness of states' compliance with their CITES obligations.[120]

If the Secretariat escalates a non-compliance issue, a Party's conformity with their treaty obligations is subject to a further tier of performance review by the Standing Committee. In addition to information provided by the Secretariat and the Party concerned,[121] the Standing Committee can gather or request further evidence from other sources, including by seeking an invitation to conduct an information-gathering and verification mission in the Party's territory or other relevant locales.[122] Thus, relative to comparable global MEAs, CITES performance-review mechanisms are 'extensive and well-functioning'.[123]

4.1.3 Multilateral non-compliance procedures

CITES, as one of the oldest MEAs, differs from many more recent MEAs that have specific treaty provisions mandating the development of a non-compliance procedure. The CITES compliance system, in contrast, reflects the cumulative evolution of secondary rules and practice over almost three decades.[124] The legal basis for the compliance system is found in four broad treaty provisions relating

117 CITES compliance procedures, above n 102, para 15.
118 CITES, CITES Compliance Matters, prepared by the Secretariat for COP 17, Doc 23, Jul 2016, para 60.
119 Raustiala, 'Reporting and Review Institutions', above n 27, 25–27.
120 J Wettestad, 'Monitoring and Verification' in D Bodansky, J Brunnée and E Hey (eds), *The Oxford Handbook of International Environmental Law* (Oxford University Press, 2007) 974, 980.
121 CITES compliance procedures, above n 102, paras 23 and 25.
122 *Ibid* para 26.
123 Raustiala, 'Reporting and Review Institutions', above n 27, 25.
124 Reeve, 'The Convention on International Trade in Endangered Species of Wild Fauna and Flora', above n 105, 136.

to Parties' annual reporting requirements,[125] the Secretariat's review and advisory functions,[126] Parties' responsibilities to report on remedial actions to the COP[127] and the COP's powers to review responses and make recommendations.[128] The formal legal basis for CITES trade sanctions is Article XIV(1)(a), which reserves the right of states to take 'stricter domestic measures regarding the conditions for trade, taking, possession or transport of specimens of species included in Appendices I, II and III, or the complete prohibition thereof'. Arguably, this provision also authorizes states' unilateral use of trade embargoes against other states,[129] which is beyond the scope of the present discussion focused on multilateral compliance processes.

The compliance procedures and practice in CITES were not codified until the release of the *Guide to CITES Compliance Procedures* in 2007.[130] The CITES compliance procedures contain non-binding[131] guidance on general principles for compliance, the compliance bodies and their tasks, the processes for handling specific compliance matters, including procedural safeguards and reporting to the COP.[132] The procedures were intended to compile and codify existing practices,[133] reflecting a move toward greater formalization of compliance processes. Due to some Parties' resistance to formally 'adopting' these procedures,[134] the COP has taken note of them and recommended that they should be referred to for compliance matters.[135] There has thus been a trend toward proceduralization in the CITES compliance system in recent decades.

4.1.4 Non-compliance response measures

The CITES compliance system utilizes a broad range of soft and intrusive non-compliance responses. The range of non-compliance response measures that have evolved through practice, and are now codified in the CITES compliance

125 CITES, above n 87, art VIII.
126 *Ibid* arts XII and XIII.
127 *Ibid* art XIII.
128 *Ibid* art XI.
129 P H Sand, 'Enforcing CITES: The Rise and Fall of Trade Sanctions' (2013) 22(3) *Review of European Community & International Environmental Law* 251, 253.
130 CITES Compliance Procedures, above n 102.
131 *Ibid* para 1. For discussion of the legality of these procedures, see Sand, 'Enforcing CITES', above n 129, 257, 258–260.
132 CITES compliance procedures, above n 102, Annex.
133 CITES, Guidelines for Compliance with the Convention, prepared by the Chairman of the Standing Committee Working Group on Compliance for COP 14, CoP14 Doc 23, 15 Jun 2000, para 6.
134 C Payne, 'Introductory Note to Convention on International Trade in Endangered Species of Wild Fauna and Flora (CITES) Compliance Procedures' (2007) 46 *International Law Materials* 1174, 1175.
135 W Wijnstekers, *The Evolution of CITES*, 9th edition (2011) 249 <https://portals.iucn.org/library/efiles/documents/CITES-012-2011.pdf>.

procedures, are: providing advice and assistance, requesting special reporting, issuing a caution, recommending specific capacity-building actions, providing in-country assistance, notifying the public of non-compliance, issuing a formal warning, requesting a compliance action plan and recommending the suspension of trade in CITES-listed species.[136] According to these procedures, trade suspensions are to be used as a measure of last resort in cases of persistent non-compliance.[137] In practice, however, trade suspension measures are frequently applied without prior recourse to less intrusive measures. As a general rule, CITES trade sanctions are country-specific, rather than species-specific, and hence have significant deterrence value by excluding sanctioned countries from *legitimate* – and lucrative – international wildlife trade.[138]

Since the late 1990s, the number of trade suspensions has increased markedly, prompting Raustiala to observe that the use of trade suspensions in CITES constitutes the most extensive reliance on 'an enforcement approach' to address non-compliance in the major global MEAs.[139] The mandate for compliance determinations is provided by numerous COP resolutions and decisions, most notably the procedures to address: (1) serious generalized non-compliance,[140] (2) Parties' failure to implement national legislation,[141] (3) non-compliance with annual reporting requirements,[142] (4) significant trade in specimens of Appendix II species,[143] (5) the protection of high-profile species such as rhinos, tigers and great apes and (6) the control of ivory trade.[144] In practice, Parties' failure to implement national legislation, identified through the National Legislation Project,[145] is the most common cause of trade-suspension recommendations.[146] The effectiveness of the CITES system of trade embargoes is evidenced by the fact that in more than 80 percent of cases, trade suspensions could be lifted within one year on the basis that the deficient performance triggering the non-compliance response had been rectified.[147]

136 CITES compliance procedures, above n 102, paras 29 and 30.
137 *Ibid* para 30.
138 P H Sand, 'Sanctions in Case of Non-Compliance and State Responsibility: *pacta sunt servanda* – Or Else?' in U Beyerlin, P-T Stoll and W Rudiger (eds), *Ensuring Compliance with Multilateral Environmental Agreements: A Dialogue Between Practitioners and Academia* (Brill Martinus Nijhoff Publishers, 2006) 259, 263.
139 Raustiala, 'Reporting and Review Institutions', above n 27, 28; see also Sand, 'Sanctions in Case of Non-Compliance and State Responsibility', above n 138, 266.
140 CITES, Compliance and Enforcement, Resolution Conf 11.3 (Rev COP 17), Oct 2016.
141 CITES, National Laws for the Implementation of the Convention, Resolution Conf 8.4 (Rev COP 15), Mar 2010.
142 CITES, National Reports, Resolution Conf 11.17 (Rev COP 17), Oct 2016.
143 CITES, Review of Significant Trade in Specimens of Appendix II Species, Resolution Conf 12.8 (Rev COP 17), Oct 2016.
144 Reeve, 'Wildlife Trade, Sanctions and Compliance', above n 115, 888.
145 CITES, National Laws for the Implementation of the Convention, above n 141.
146 Sand, 'Enforcing CITES', above n 129, 254–256.
147 *Ibid* 255.

5 Similarities and differences between the compliance systems

There are key similarities and differences between the three selected compliance systems, which underscore their suitability as case studies for evaluating the benefits of proceduralization. As mentioned in Chapter 1, each compliance system relies on rules and decisions of an administrative character, and involves state actors as the creators and enforcers, and subjects and agents, of administrative decisions. Furthermore, as this chapter has shown, they each have a similar hierarchical structure, in which there are a number of intermediary steps between national reporting and, where necessary, the determination of consequences for non-compliance. As each compliance system represents a multi-tiered regulatory framework, systemic tendencies can be brought into relief, allowing conclusions to be drawn about the net gains associated with proceduralization.[148]

In addition to these practical considerations, there are theoretical considerations that support a focus on the three selected compliance systems. This section applies matrices developed by Esty to demonstrate that the chosen case studies are similar, and different from other global MEAs, as their institutional design and the particular issues that they address make them particularly likely to raise legitimacy questions. The adoption of administrative procedures is one way to address these concerns.[149] This assumes that proceduralization enhances normative legitimacy, which invokes notions of an institution's 'right to rule', and a subject's correlative 'obligation to obey'.[150] The link that Esty draws between right process and legitimacy is consistent with other scholars' claims that procedural rigor and integrity can enhance legitimacy in international law.[151] On a descriptive level, whether compliance systems' rules and decisions *do* in fact exert a compliance pull on states in relation to their multilateral environmental commitments is ultimately an empirical question that is yet to be tested.[152] Despite this, for present purposes, applying a normative legitimacy lens is instructive as it underscores that the selected case studies in principle have the most to gain of any MEA compliance system from adopting administrative procedures.

148 See Chapter 7, section 2.

149 Esty, 'Good Governance at the Supranational Level', above n 1, 1509–1514. Esty argues that 'the procedural rigor of administrative law is a critical tool for refining international governance and legitimizing the exercise of supranational authority': *ibid* 1495.

150 D Bodansky, 'Legitimacy' in D Bodansky, J Brunnée and E Hey (eds), *The Oxford Handbook of International Environmental Law* (Oxford University Press, 2007) 704, 708.

151 See, eg, D Hovell, *The Power of Process: The Value of Due Process in Security Council Sanctions Decision-Making* (Oxford University Press, 2016) 61–63; T Franck, *Fairness in International Law and Institutions* (Oxford University Press, 1995) 2; S Cassese, 'A Global Due Process of Law?' in G Anthony, J-B Auby, J Morison and T Zwart (eds), *Values in Global Administrative Law* (Hart Publishing, 2011) 17, 19–20.

152 D Bodansky, 'Legitimacy in International Law and International Relations' in J L Dunoff and M A Pollack (eds), *Interdisciplinary Perspectives on International Law and International Relations: The State of the Art* (Cambridge University Press, 2012) 321, 332–334.

Esty proposes two matrices representing the interplay between institutional factors, and the type of issue a global decision-making body is addressing, to explain the scenarios in which the legitimacy of global regulatory bodies is most likely to be drawn into question.[153] Each of these matrices is applied in turn below. First, in relation to the institutional determinants of legitimacy, Esty proposes that issues related to the 'depth or "thickness" of governance' can be arrayed on a matrix. The horizontal axis represents a continuum of 'soft/informal policymaking' to 'hard/formal policymaking', and the vertical axis represents a continuum of 'intergovernmental decisionmaking' to 'supranational decisionmaking'.[154]

The horizontal axis represents how formal and binding international rules and decisions are.[155] As previously mentioned, the formal legal status of the administrative decision making in MEA compliance systems is disputed but, for a variety of reasons, states tend to commit to and honor such decisions.[156] Moreover, as Koskenniemi notes, characterizing treaty violations as 'non-compliance' rather than breaches of international law that are potentially subject to principles of state responsibility for wrongful acts signifies that environmental obligations are '"softer" than other – "normal" – obligations'.[157] The disputed legal formality and binding force of the rules and decisions within MEA compliance systems would tend to indicate that all such determinations fall toward the 'soft/informal policy making'[158] end of this continuum.

However, as Brunnée observes, a 'hardening' of MEAs' approaches to compliance can be achieved through 'soft' law mechanisms.[159] This perspective suggests that there are factors affecting the 'binding effect' or 'strength' of IEL that transcend the soft versus hard law dichotomy.[160] I argue that the degree of 'intrusiveness'[161] of the potential consequences for non-compliance reflects this hardening process, and is an important point of distinction between MEAs. Specifically, the compliance systems with the most punitive potential consequences for non-compliance, which are tied to trade suspensions and the deprivation of other rights and privileges under the treaty in question, are considerably more intrusive for states than their facilitative counterparts.

153 *Ibid.*

154 Esty, 'Good Governance at the Supranational Level', above n 1, 1509.

155 *Ibid.*

156 T Gehring, 'Treaty-Making and Treaty Evolution' in D Bodansky, J Brunnée and E Hey (eds), *The Oxford Handbook of International Environmental Law* (Oxford University Press, 2007) 467, 492–495.

157 M Koskenniemi, 'Breach of Treaty or Non-Compliance? Reflections on the Enforcement of the Montreal Protocol' (1992) 3(1) *Yearbook of International Environmental Law* 123, 127.

158 Esty, 'Good Governance at the Supranational Level', above n 1, 1509.

159 J Brunnée, 'Multilateral Environmental Agreements and the Compliance Continuum' in G Winter (ed), *Multilevel Governance of Global Environmental Change: Perspectives from Science, Sociology and the Law* (Cambridge University Press, 2006) 387, 407.

160 *Ibid.*

161 *Ibid* 403.

The Montreal, Kyoto and CITES compliance systems are similar – and distinctive vis-à-vis other global MEA compliance systems – in terms of the intrusiveness of the potential consequences for non-compliance. Most MEA compliance systems employ facilitative measures such as the provision of advice and financial and technical assistance,[162] which recognizes that non-compliance is often attributable to Parties' technical, financial and regulatory capacity constraints rather than deliberate contravention.[163] Slightly stronger measures include requests for a Party to undertake additional reporting or produce a compliance action plan. As a further step, 'censure' measures such as the issuance of cautionary statements, publication of cautions on the regime website and declarations of non-compliance can be employed. These censure measures reflect an understanding that failure to comply with international law obligations has a reputational cost, which may damage other states' perceptions of a particular country's commitment to long-term international relationships premised on mutual co-operation.[164] All of these types of measures are best described as 'soft' and relatively non-intrusive.

The strongest set of measures that only the Montreal, Kyoto and CITES compliance systems authorize are suspensions of trade and removal of non-compliant Parties' rights and privileges under the regime.[165] As previously mentioned, in the Montreal compliance system, trade restrictions were controversially imposed on certain countries with economies in transition in the mid-1990s.[166] Under the Kyoto compliance system, the suspension of Parties' eligibility to participate in the Protocol's flexibility mechanisms is a type of economic sanction that was relatively frequently recommended as a consequence of non-compliance in the first commitment period.[167] In the case of CITES, suspensions of trade in CITES-listed species may be recommended for various infractions, and have been used extensively in practice.[168] These non-compliance responses are intrusive as they have economic

162 F Romanin Jacur, 'Controlling and Assisting Compliance: Financial Aspects' in T Treves et al (eds), *Non-Compliance Procedures and Mechanisms and the Effectiveness of International Environmental Agreements* (TMC Asser Press, 2009) 419.

163 See Chapter 1, section 2.

164 See, eg, A T Guzman, 'A Compliance-Based Theory of International Law' (2002) 90 *California Law Review* 1823, 1884; A T Guzman, *How International Law Works: A Rational Choice Theory* (Oxford University Press, 2008) 33–34.

165 J Bulmer, 'Compliance Regimes in Multilateral Environmental Agreements' in J Brunnée, M Doelle and L Rajamani (eds), *Promoting Compliance in an Evolving Climate Regime* (Cambridge University Press, 2012) 55, 72–73; G Ulfstein, 'Dispute Resolution, Compliance Control and Enforcement in International Environmental Law' in G Ulfstein, T Marauhn and A Zimmermann (eds), *Making Treaties Work: Human Rights, Environment and Arms Control* (Cambridge University Press, 2007) 115, 130.

166 See above n 50.

167 A Zahar, 'A Bottom-Up Compliance Mechanism for the Paris Agreement', *Chinese Journal of Environmental Law* (forthcoming, 2017) 9 <https://papers.ssrn.com/sol3/papers.cfm?abstract_id=2912560>.

168 There has been debate about the legality of such measures, including in relation to their compatibility with the Vienna Convention on the Law of Treaties: *Vienna Convention on*

implications for states. Procedural rigor in the making of compliance decisions is desirable to justify the imposition of punitive measures.[169]

The intersecting vertical spectrum in Esty's first matrix represents a continuum reflecting the locus of decision-making authority.[170] In this regard, too, the selected case studies are similar. According to Esty, decision making along this spectrum ranges from intergovernmental decision making to supranational decision making, with the latter having a greater need to demonstrate due process to legitimate its decisions.[171] In MEA compliance systems, considerable authority to recommend measures in instances of non-compliance has been delegated to subsidiary bodies – the Implementation Committee in the Montreal Protocol, the Enforcement Branch in the Kyoto Protocol and the Standing Committee in CITES – rather than being vested in the respective Conference of the Parties as an intergovernmental decision-making forum. Accordingly, these bodies may have a greater need to demonstrate that their exercise of the powers delegated to them is legitimate, especially in the light of the stronger and more intrusive measures, outlined above, that are possible consequences of non-compliance in these three MEAs. For example, the Russian Federation has questioned the Implementation Committee's authority to recommend trade restrictions under the Montreal Protocol,[172] Canada has expressed its displeasure regarding actions taken by the Enforcement Branch in the Kyoto Protocol[173] and the legal basis of trade suspensions under CITES has been challenged.[174] Therefore, according to Esty's matrix, the types of authority exercised by these compliance committees fall closer toward the supranational decision-making end of the continuum.[175]

For these reasons, the three case study MEA compliance systems, which have intrusive consequences for non-compliance and high reliance on supranational decision making, fall in the upper-right quadrant of Esty's matrix, as is depicted in the following diagram:

the Law of Treaties, opened for signature 23 May 1969, 1155 UNTS 331 (entered into force 27 January 1980); Bulmer, 'Compliance Regimes in Multilateral Environmental Agreements', above n 165, 72. On the topic of the legality of the suspension of rights and privileges under the Montreal Protocol's non-compliance procedure, see Koskenniemi, 'Breach of Treaty or Non-Compliance?', above n 157, 146, 152–155.

169 Esty, 'Good Governance at the Supranational Level', above n 1, 1509–1514; Ulfstein, 'Dispute Resolution, Compliance Control and Enforcement in International Environmental Law', above n 165, 128.

170 Esty, 'Good Governance at the Supranational Level', above n 1, 1509.

171 *Ibid.*

172 UNEP, Report of the Seventh Meeting of the Parties to the Montreal Protocol on Substances that Deplete the Ozone Layer, UNEP/OzL.Pro.7/12, 27 Dec 1995, para 44. See further Chapter 4, section 3.

173 M Doelle, 'Experience with the Facilitative and Enforcement Branches of the Kyoto Compliance System' in J Brunnée, M Doelle and L Rajamani (eds), *Promoting Compliance in an Evolving Climate Regime* (Cambridge University Press, 2012) 102, 114–115.

174 Sand, 'Enforcing CITES', above n 129, 257.

175 Esty, 'Good Governance at the Supranational Level', above n 1, 1509.

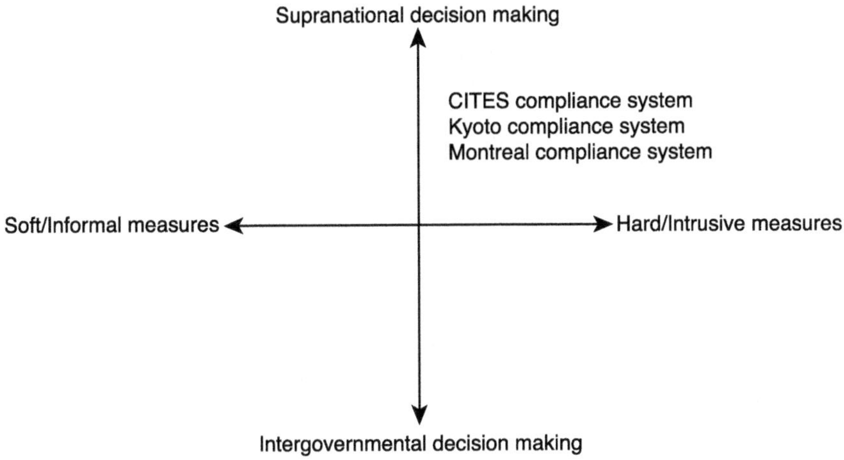

Figure 2: Institutional factors contributing to legitimacy concerns (adapted from Esty, 'Good Governance at the Supranational Level', above n 1, 1509)

To paraphrase Esty, '[i]n this zone, the need for an administrative law regime intensifies because the authority exercised by supranational institutions can significantly limit the actions of, [and impact upon] domestic governments'.[176] Accordingly, the adoption of administrative procedures is highly desirable in these settings.

The application of Esty's second matrix illuminates potential differences between MEAs regarding the extent of interdependence required to fulfill their mandates, and the level of contestation that the issue they are addressing attracts. This matrix maps the costs and benefits of global governance along two intersecting spectrums: the vertical axis represents a scale of international interdependence for varying types of issues, ranging from 'limited interdependence' to 'deep interdependence',[177] and the intersecting horizontal axis represents a continuum of issue politicization, ranging from 'scientific/technical issues' at one end, to 'political/value-laden issues' at the other.[178]

Prima facie, it appears that all global MEAs, which were negotiated to facilitate collective action on issues of international environmental concern, are likely to require deep interdependence to achieve their goals. In this regard, global MEAs can be contrasted with bilateral environmental agreements and water pollution control agreements, for example, which would be clustered toward the limited interdependence end of the continuum.[179] However, a closer examination reveals that there are in fact variations between global MEAs in terms of the extent of

176 *Ibid* 1510.
177 *Ibid* 1512.
178 *Ibid.*
179 *Ibid* 1512–1513.

interconnectedness and co-operation required of states to implement their treaty obligations. For example, the Ramsar Convention[180] places responsibility upon Parties to list and protect significant wetland sites by establishing nature reserves, using sites wisely and encouraging the growth in waterfowl populations.[181] However, as there are no legal penalties for non-compliance, a country can achieve nominal compliance by simply listing a wetland site and making tokenistic protection efforts.[182] Moreover, there are national environmental and tourism benefits that may accrue from listing sites and taking protection measures even in the absence of international co-operation. In such instances, the extent to which states must depart from what they may ordinarily have done in the absence of the treaty is likely to be comparatively limited.[183]

In contrast, reducing global levels of ozone-depleting substances and greenhouse gas emissions under the Montreal Protocol and the Kyoto Protocol, respectively, and regulating international trade in endangered species under CITES, represent a greater departure from business as usual. This is because such endeavors provide fewer direct national benefits to the Parties and require substantial efforts to achieve international co-operation. Compliance theorists aligned with the enforcement model claim that it is no coincidence that the types of MEAs with the most intrusive consequences for non-compliance are also the agreements requiring deeper co-operation for their implementation – indeed, they argue that the former is necessary to secure the latter in situations where strong incentives for non-compliance exist.[184] As noted above, the consequences of non-compliance in the Montreal, Kyoto and CITES compliance systems are potentially more intrusive than in other MEAs, reinforcing the argument that deeper interdependence is required in these regulatory settings.

Along the intersecting horizontal spectrum of Esty's second matrix, there are notable differences between the three selected case studies in terms of the extent to which the environmental issue being addressed is politically charged and underpinned by scientific agreement. Esty suggests that the Montreal Protocol

180 *Convention on Wetlands of International Importance Especially as Waterfowl Habitat*, opened for signature 2 February 1971, 11 ILM 963 (entered into force 21 December 1975) ('Ramsar Convention').

181 Raustiala, 'Reporting and Review Institutions', above n 27, 17.

182 D G Victor, 'Enforcing International Law: Implications for an Effective Global Warming Regime' (1999) 10 *Duke Environmental Law and Policy Forum* 147, 155.

183 G W Downs, D M Rocke and P N Barsoom, 'Is the Good News about Compliance Good News about Cooperation?' (1996) 50(3) *International Organization* 379, 380.

184 *Ibid* 387. In the context of examining a range of international institutions and agreements, Downs, Rocke and Barsoom argue that the Montreal Protocol does not represent an example of deep co-operation: *ibid* 391. Cf Victor, who contends that the stronger measures for non-compliance under the Montreal Protocol *do* provide evidence to support the enforcement model: Victor, 'Enforcing International Law', above n 182, 164–167. Consistently with Victor's logic, my argument is that among the global MEAs, the Montreal Protocol, as well as the Kyoto Protocol and CITES, do require deeper co-operation than other MEAs.

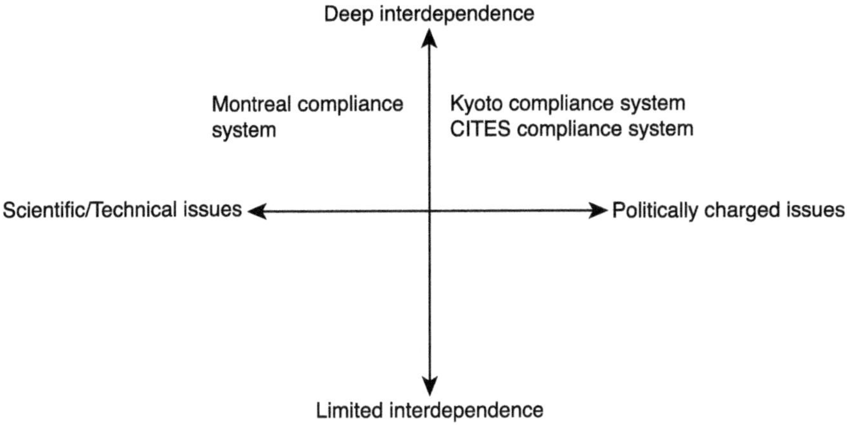

Figure 3: Issue areas contributing to legitimacy concerns (adapted from Esty, 'Good Governance at the Supranational Level', above n 1, 1512)

falls in the upper-left quadrant of his second matrix, as the degree of interdependence required to phase out ozone-depleting substances is 'substantial', while the 'issues are relatively narrow and technical'.[185] In contrast, the issues that are the focus of the Kyoto Protocol and CITES are significantly more politicized. In relation to the former, the issues of climate-change mitigation and adaptation are characterized by deep North–South tensions.[186] Moreover, climate-change science is contested in some political contexts, and there are divergent values and approaches to achieving mitigation and adaptation across countries. Similarly, the issue of restricting trade under CITES is politicized along North–South fault lines, especially in relation to developing range states' sustainable use of endangered species for livelihood purposes.[187] Hence, the Kyoto Protocol and CITES provide examples of global MEAs characterized by the need for deep interdependence on an issue that is highly charged politically, operating in the upper-right quadrant of Esty's matrix. Significantly, Esty observes that '[o]perating in this zone of greater political sensitivity requires more fully developed procedures to establish the legitimacy of policymaking'.[188]

In addition to these similarities, there are some salient differences in regulatory and procedural design between the three compliance systems, which provide relevant variation for this analysis. As foreshadowed in Chapter 1, each of these compliance systems has particular features that are conducive to, or limit, the benefits of proceduralization. The Montreal compliance system differs from the other two case studies with respect to the extent of reliance on conditional funding as a regulatory

185 Esty, 'Good Governance at the Supranational Level', above n 1, 1513.
186 See further Chapter 5.
187 See further Chapter 6.
188 Esty, 'Good Governance at the Supranational Level', above n 1, 1513.

technique to buttress procedural regard opportunities. The Kyoto compliance system is highly proceduralized and technical experts play a more influential role in decision making than in the compliance systems under the Montreal Protocol and CITES. In CITES, there has been a shift in the dominant regime paradigm from an ostensible consensus on preservation to a greater accommodation of sustainable use perspectives, which signifies greater legal and regulatory pluralism in conservation discourses. Moreover, creating opportunities for participation of civil-society representatives provides an additional mechanism for bringing to the fore the interests of states affected by CITES compliance decisions. As the analysis in Chapters 4–6 highlights, each of these differences between the regimes indicates conditions that promote the potential for administrative procedures to constrain power and enhance regard for affected interests.

6 Conclusion

This overview of the three selected MEA compliance systems underscores their suitability as case studies for this book. Each compliance system represents a complex administrative apparatus, characterized by states' provision of performance-review information, multilateral non-compliance procedures and non-compliance response measures. The Montreal, Kyoto and CITES compliance systems are similar – and diverge from other global MEAs – as they are likely to attract legitimacy concerns due to their institutional design and politically sensitive subject matters. Therefore, to the extent that adopting administrative procedures enhances legitimacy, these systems have the most to gain from doing so. In the following case study chapters, the presence and effect of mechanisms for accountability, transparency, participation, reason giving and review are analyzed across these multi-tiered compliance frameworks, bringing into relief the similarities and differences in the roles played by procedure in each system.

4 The Montreal compliance system

An exemplar of procedural regard

1 Introduction

The Montreal compliance system provides an apposite first case study for this book as it exemplifies the roles that administrative procedures can play in promoting procedural regard for the interests of affected states. The Montreal Protocol is regarded as a forerunner in terms of practically embedding the principle of common but differentiated responsibilities (CBDR) in multilateral environmental agreements (MEAs),[1] which underpinned the development of a compliance framework that is well attuned to less powerful states' compliance interests. From the outset, the Montreal Protocol differentiated between developing states' ('Article 5 Parties') and developed states' ('non-Article 5 Parties') commitments. While industrialized states were expected to take the lead in reducing their production and consumption of chlorofluorocarbons (CFCs) and other ozone-depleting substances such as halons, developing countries were granted a grace period in which they could temporarily *increase* their use of ozone-depleting substances before assuming their reduction commitments.[2] Moreover, provision was made for the transfer of technology to help phase out ozone-depleting substances, and financial mechanisms to assist developing countries and later countries with economies in transition (CEITs) to shift to ozone-friendly technology.[3] Procedural accountability arrangements that developed within the Montreal compliance system conditionally linked these incentives to developing and

1 U Beyerlin and T Marauhn, *International Environmental Law* (Hart Publishing, 2011) 156.
2 *Ibid.* For example, developing countries were given a staggered timetable allowing them an extra ten years – that is, until 2006 – to phase out the most noxious ozone-depleting substances: D G Victor, 'Enforcing International Law: Implications for an Effective Global Warming Regime' (1999) 10 *Duke Environmental Law and Policy Forum* 147, 161.
3 *Montreal Protocol on Substances that Deplete the Ozone Layer 1987*, opened for signature 16 September 1987, 1989 UNTS 28 (entered into force 1 January 1989) art 10 (financial mechanism) and art 10A (technology transfer); K Madhava Sarma and K N Taddonio, 'The Role of Financial Assistance by the Multilateral Fund in Technology Change to Protect the Ozone Layer' in Z Cerefos, G Contopoulos and G Skalkeas (eds), *Twenty Years of Ozone Decline: Proceedings of the Symposium for the 20th Anniversary of the Montreal Protocol* (Springer, 2009) 441, 444.

transition countries' adherence to selected compliance benchmarks agreed between the Party concerned and the Implementation Committee.[4]

The privileging of the principle of CBDR, and the establishment of mechanisms for financial and technical assistance for developing countries in the text of the Montreal Protocol, created a framework for procedural and substantive regard for less powerful states' interests in this compliance system. As outlined in Chapter 2, procedural regard is evidenced when decision makers take into account the situation and interests of countries affected by their decisions, whereas substantive regard implies outcomes that are favorable to weaker actors.[5] The considerable overlap between the two is acknowledged, however this chapter is primarily concerned with the former – whether and how administrative procedures have been employed in ways that take account of the interests of affected states in compliance decision making.

The application of the evaluative framework for proceduralization to this compliance system reveals three key trends. First, there is evidence of the adoption of a limited set of administrative procedures for accountability, transparency, participation and reason giving, but avenues for review are circumscribed. Second, decision making in the compliance system has become increasingly procedurally constrained, leading to the emergence of a consistent pattern of accountability arrangements to respond to non-compliance, and facilitate non-compliant Parties' return to compliance. However, these gains have been undermined by poor data quality and Parties' expectations of negotiable compliance standards. Third, in many respects, the compliance procedures developed under the Montreal Protocol exemplify administrative procedures that facilitate regard for the interests of developing and transition countries. Conditional funding has proven to be a powerful tool for buttressing and reinforcing procedural regard opportunities. This thematic analysis is crystallized by a vignette of the non-compliance proceedings involving the Russian Federation in section 3, which demonstrates the extension of procedural regard to a country in transition.

2 Evaluating proceduralization in the Montreal compliance system

This section applies the evaluative framework articulated in Chapter 2 to the Montreal compliance system. A historical perspective is adopted to assess the degree of adjustment toward, or away from, proceduralization over time. As there is minimal scope for review in the Montreal compliance system, this analysis is chiefly concerned with procedures for accountability, transparency, participation and reason giving.

4 G M Bankobeza, *Ozone Protection: The International Legal Regime* (Eleven International Publishing, 2005) 232.

5 R B Stewart, 'Remedying Disregard in Global Regulatory Governance: Accountability, Participation, and Responsiveness' (2014) 108 *American Journal of International Law* 211, 224. See Chapter 2, section 5.

2.1 Accountability

Despite an ongoing commitment to flexibility, the evolution of accountability procedures in the Montreal compliance system reflects increasingly procedurally constrained decision making, and regard for affected states' interests. The Implementation Committee represents the locus of compliance decision making within the Montreal compliance system,[6] and is thus the primary focus of this discussion on accountability. Some elements of accountability are satisfied in the Implementation Committee – namely, states are required to justify their conduct, and are answerable to the Committee for deficient performance.[7] In relation to consequences for conduct that does not meet specified standards, the standards themselves are treated as flexible and revisable, which indicates a departure from administrative decision-making ideals. Nonetheless, a consistent pattern of accountability arrangements to facilitate non-compliant Parties' return to compliance has emerged.[8] In particular, there is evidence over time of a greater prioritizing of the equitable treatment of Parties in like circumstances alongside regard for the individual circumstances of each country,[9] which is consistent with an administrative style of decision making. Thus, the Montreal compliance system has created zones for compliance decision making that allow for a degree of discretionary leeway in the light of national circumstances, yet are constrained by administrative mechanisms.

2.1.1 Accountability and self-imposed targets

In its early years of operation, the Implementation Committee was primarily concerned with data-reporting problems, including in relation to adequate reporting of baseline data.[10] The grace period for developing countries ended in 1999, resulting in a significant increase in the number of Parties in substantive

6 See Chapter 3, section 2.1.1.

7 Stewart, 'Remedying Disregard in Global Regulatory Governance', above n 5, 253.

8 Y Shigeta, *International Judicial Control of Environmental Protection: Standard Setting, Compliance Control and the Development of International Environmental Law by the International Judiciary* (Wolters Kluwer, 2010) 120.

9 A Cardesa-Salzmann, 'Constitutionalising Secondary Rules in Global Environmental Regimes: Non-Compliance Procedures and the Enforcement of Multilateral Environmental Agreements' (2011) 24(1) *Journal of Environmental Law* 103, 131–132. See generally, Ozone Secretariat, *Implementation Committee under the Non-Compliance Procedure of the Montreal Protocol on Substances that Deplete the Ozone Layer: Primer for Members* (October 2007) 4 <http://ozone.unep.org/new_site/en/resources.php>.

10 D G Victor, 'The Operation and Effectiveness of the Montreal Protocol's Non-Compliance Procedure' in D G Victor, K Raustiala and E B Skolnikoff (eds), *The Implementation and Effectiveness of International Environmental Commitments* (MIT Press, 1998) 137, 143; A Gillespie, 'Implementation and Compliance Concerns in International Environmental Law: The State of the Art within Three International Regimes' (2003) 7 *New Zealand Journal of Environmental Law* 53, 70.

non-compliance. To date, however, no developing country has been deprived of assistance or had its rights and privileges suspended.[11] The Implementation Committee has consistently adopted a co-operative, flexible and facilitative approach to compliance,[12] which provides a favorable context for procedural regard.

The Implementation Committee's response to supporting Parties' return to compliance has become increasingly proceduralized over time. One notable example of this is the procedures, forged with the help of the Implementation Committee, that were developed to increase developing countries' accountability for the financial assistance they receive to facilitate their implementation of, and compliance with, the Montreal Protocol.[13] The Montreal Protocol's financial mechanism, by which developed countries' donations are disbursed to support developing countries to meet their reporting and phase-out requirements, was introduced in 1991. Ongoing concerns about some Parties' failure to meet their reporting requirements prompted the Implementation Committee to agree in 1998 on a process for extending conditional funding arrangements to Parties that are found to be in non-compliance.[14] This means that a non-compliant Party's continuing receipt of funding assistance from either the Multilateral Fund or the Global Environment Facility hinges upon meeting benchmarks the Implementation Committee selects from the Party's compliance action plan.[15] Parties that meet these agreed commitments are treated as remaining 'in good standing' and receive favorable consideration when applying for financial assistance. Parties that deviate from these commitments are threatened with punitive measures, such as deprivation of funding assistance and export restrictions.[16]

The use of specified benchmarks against which Parties' progress toward full compliance can be assessed reflects a procedural accountability arrangement in which the capacity constraints of individual countries are considered in compliance determinations. These arrangements are consistent with procedural regard. However, it is notable that the procedurally constrained political negotiation practices that occur in the compliance system at times actively accommodate departure from binding obligations in the Montreal Protocol as states are

11 Ozone Secretariat, *Handbook for the Montreal Protocol on Substances that Deplete the Ozone Layer, Tenth Edition* (2016) <http://ozone.unep.org/sites/ozone/files/Publications/Handbooks/MP-Handbook-2016-English.pdf>.

12 Shigeta, *International Judicial Control of Environmental Protection,* above n 8, 120; Bankobeza, *Ozone Protection*, above n 4, 261.

13 Shigeta, *International Judicial Control of Environmental Protection*, above n 8, 131–132; D G Victor, 'The Early Operation and Effectiveness of the Montreal Protocol's Non-Compliance Procedure' (1996) 12 <http://irps.ucsd.edu/dgvictor/publications/Victor_Article_1996_Early%20Operation%20and%20Effectiveness.pdf>.

14 UNEP, Report of the Implementation Committee under the Non-Compliance Procedure for the Montreal Protocol on the Work of its Twentieth Meeting, UNEP/OzL.Pro/ImpCom/20/4, 9 Jul 1998, paras 24–33.

15 *Ibid* para 32.

16 *Ibid* para 33.

accountable for pre-agreed and standardized commitments,[17] rather than their treaty commitments. These accountability arrangements based on individually tailored compliance action plans agreed between the Party concerned and the Implementation Committee support Klabbers' observation that compliance is at times 'subject to negotiations'.[18] Such arrangements stand in stark contrast to the traditional rules of state responsibility for breach of an international obligation under general international law, raising questions of legitimacy from a traditional international law perspective.[19] Despite this, the striking successes of the Montreal Protocol and its compliance system[20] highlight possibilities for procedural arrangements that promote and facilitate compliance without necessarily insisting on strict adherence to binding treaty rules.[21]

2.1.2 Individually tailored versus predictable consequences of non-compliance

In relation to the types and consequences of non-compliance, the Parties to the Montreal Protocol have favored a discretionary, flexible approach[22] that is not procedurally prescriptive. An 'Indicative List of Possible Situations of Non-Compliance' was proposed at the third MOP in 1991,[23] but not adopted at the fourth MOP in 1992.[24] While Parties agreed that the list could encompass non-compliance with provisions pertaining to control measures, control of trade with non-Parties and non-reporting, there was disagreement on the obligatory nature of financial contributions to the Multilateral Fund. In the absence of an ability to reach consensus on a comprehensive list, the Parties opted not to adopt a list at all.[25]

17 M Koskenniemi, 'Breach of Treaty or Non-Compliance? Reflections on the Enforcement of the Montreal Protocol' (1992) 3(1) *Yearbook of International Environmental Law* 123, 162; J Klabbers, 'Compliance Procedures' in D Bodansky, J Brunnée and E Hey (eds), *The Oxford Handbook of International Environmental Law* (Oxford University Press, 2007) 995, 996–998, 1007–1009; Bankobeza, *Ozone Protection*, above n 4, 300.

18 Klabbers, 'Compliance Procedures', above n 17, 1001; Koskenniemi, 'Breach of Treaty or Non-Compliance?', above n 17, 147.

19 Koskenniemi, 'Breach of Treaty or Non-Compliance?', above n 17, 162.

20 Cardesa-Salzmann, 'Constitutionalising Secondary Rules in Global Environmental Regimes', above n 9, 103. See further Chapter 3, section 2.

21 See further Chapter 7, section 2.3. See also A Huggins, 'Administrative Procedures and Rule of Law Values in the Montreal Compliance System' in Christina Voigt (ed), *The Environment in International Courts and Tribunals: Questions of Legitimacy* (Cambridge University Press, forthcoming).

22 S Borras, 'Comparative Analysis of Selected Compliance Procedures under Multilateral Environmental Agreements' in S Maljean-Dubois and L Rajamani (eds), *Implementation of International Environmental Law* (Martinus Nijhoff Publishers, 2011) 319, 348.

23 UNEP, Decision III/2: Non-Compliance Procedure, UNEP/OzL.Pro.3/2, 21 Jun 1991.

24 UNEP, Report of the Fourth Meeting of the Parties, UNEP/OzL.Pro.4/15, 25 Nov 1992.

25 Borras, 'Comparative Analysis of Selected Compliance Procedures', above n 22, 346; UNEP, Report of the Fourth Meeting of the Parties, above n 24.

Furthermore, during the 1998 review of the non-compliance procedure,[26] a proposal for guidance to be provided to assist the Implementation Committee in matching non-compliance responses to particular types of non-compliance was considered. This proposal was ultimately rejected in favor of a continued reliance on 'discretion to adapt their response to the particular case'.[27]

Despite this resistance to procedural guidelines regarding the consequences that should attach to different types of non-compliance, accountability practices in the Montreal compliance system have in fact been relatively consistent and predictable. As noted above, a common practice of making the continued receipt of financial assistance contingent upon reaching time-bound benchmarks specified in compliance action plans emerged for developing countries in the early 1990s. Then, in the mid-1990s, the non-compliance proceedings against Belarus, Bulgaria, Poland, Russia and Ukraine extended a variation of this archetypal response to CEITs, creating a clear precedent for subsequent proceedings against CEITs in 1998 and 1999.[28] The process adopted in these cases involved three elements: (1) a detailed acknowledgement and explanation of the non-compliance, (2) a statement that a Party in non-compliance with its ozone-depleting substances obligations under the Protocol will remain a 'member in good standing' that is eligible for international financial assistance so long as compliance benchmarks are met and (3) a warning regarding the potential punitive measures that may be

26 UNEP, Decision IV/5: Non-Compliance Procedure, UNEP/OzL.Pro.4/1, 25 Nov 1992 ('Non-compliance procedure').

27 UNEP, Report on the Work of the Ad Hoc Working Group of Legal and Technical Experts on Non-compliance with the Montreal Protocol, UNEP/OzL.Pro/WG.4/1/3, 18 Nov 1998, para 40. During its non-compliance proceedings in the mid-1990s, Russia proposed the need for clearer criteria to facilitate the Implementation Committee's exercise of objective judgement as to whether the cause of non-compliance was a result of willful breach or factors beyond the control of the Party concerned, but such criteria were not adopted.

28 Specifically, the proceedings against Azerbaijan, Belarus, the Czech Republic, Estonia, Latvia, Lithuania, the Russian Federation, Ukraine and Uzbekistan in 1998, and Bulgaria and Turkmenistan in 1999: UNEP, Decision X/20: Compliance with the Montreal Protocol by Azerbaijan, UNEP/OzL.Pro.10/9, 3 Dec 1998; UNEP, Decision X/21: Compliance with the Montreal Protocol by Belarus, UNEP/OzL.Pro.10/9, 3 Dec 1998; UNEP, Decision X/22: Compliance with the Montreal Protocol by the Czech Republic, UNEP/OzL.Pro.10/9, 3 Dec 1998; UNEP, Decision X/23: Compliance with the Montreal Protocol by Estonia, UNEP/OzL.Pro.10/9, 3 Dec 1998; UNEP, Decision X/24: Compliance with the Montreal Protocol by Latvia, UNEP/OzL.Pro.10/9, 3 Dec 1998; UNEP, Decision X/25: Compliance with the Montreal Protocol by Lithuania, UNEP/OzL.Pro.10/9, 3 Dec 1998; UNEP, Decision X/26: Compliance with the Montreal Protocol by the Russian Federation, UNEP/OzL.Pro.10/9, 3 Dec 1998; UNEP, Decision X/27: Compliance with the Montreal Protocol by Ukraine, UNEP/OzL.Pro.10/9, 3 Dec 1998; UNEP, Decision X/28: Compliance with the Montreal Protocol by Uzbekistan, UNEP/OzL.Pro.10/9, 3 Dec 1998; UNEP, Decision XI/24: Compliance with the Montreal Protocol by Bulgaria, UNEP/OzL.Pro.11/10, 17 Dec 1999; UNEP, Decision XI/25: Compliance with the Montreal Protocol by Turkmenistan, UNEP/OzL.Pro.11/10, 17 Dec 1999.

applied in cases of persistent non-compliance.[29] This type of precedent-based approach, which has applied to both developing countries and CEITs, aligns with administrative decision-making ideals of consistency and predictability. The conditional funding arrangements also take into account the capacity constraints experienced by these states, which is consonant with procedural regard.

Hence, the 'basic posture'[30] of recent practices in the Implementation Committee reflects an administrative style of decision making that demonstrates regard for the merits of individual cases. This approach is reinforced in the *Primer for Members* produced by the Ozone Secretariat for Implementation Committee members in 2007, which lists sixteen 'routine procedural non-compliance matters' and a set of standardized recommendations that the Committee has agreed should apply for these types of matters. In such cases, the draft recommendations prepared by the Secretariat for the consideration of the Committee are based on text approved by the Committee at previous meetings to deal with similar cases, and can therefore be approved without extended deliberation or fact-finding.[31] Recent practice evidences a mixed approach: in some instances, the Implementation Committee undertakes individual review to obtain an in-depth understanding of the particular circumstances of the non-compliant Party, and in other instances a 'blanket approval' process based on previous precedents is used, promoting the like treatment of analogous cases.[32] This signals a gradual shift from privileging an approach tailored to the individual circumstances of each country, to one that balances this imperative with the equitable treatment of Parties in like circumstances according to pre-determined standards. This latter trend is consistent with administrative decision-making modes and procedural regard.

2.2 Transparency

2.2.1 Internal and external transparency

Despite a commitment to safeguarding confidentiality, the Montreal compliance system has employed numerous mechanisms for promoting internal and external transparency. However, one critical issue that has hampered the realization of both transparency and accountability aims is poor data quality[33] in states' national reports, which form the foundation for subsequent compliance decisions. The expectation of deference to the information reported by sovereign states, without

29 *Ibid*; A Gillespie, *Climate Change, Ozone Depletion, and Air Pollution: Legal Commentaries with Policy and Scientific Considerations* (Martinus Nijhoff Publishers, 2006) 221–226.

30 P Nonet and P Selznick, *Toward Responsive Law: Law & Society in Transition* (Transaction Publishers, 2001) 17.

31 Ozone Secretariat, *Primer for Members*, above n 9, 45–58.

32 *Ibid* 14.

33 D G Victor, *Global Warming Gridlock: Creating More Effective Strategies for Protecting the Planet* (Cambridge University Press, 2011) 230.

any independent verification process, reflects a traditional diplomatic mindset, and does not prioritize robust performance-review information.

Protecting states' confidentiality is a primary concern in the Montreal compliance system. States' reported data on ozone-depleting substances is treated as confidential due to concerns about industrial secrecy and competitiveness.[34] The meetings of both the Implementation Committee and the MOP are closed, unless the meeting decides otherwise.[35] The non-compliance procedure urges those involved in Implementation Committee deliberations to protect the confidentiality of information received in confidence.[36] These requirements align with diplomatic norms of secrecy,[37] rather than the broader transparency ideals in contemporary global environmental governance.[38]

Nonetheless, a degree of internal and external transparency is promoted by various procedural requirements, and current practices frequently exceed these minimum expectations. For example, the Implementation Committee reports are to be made available to any person on request, and the information exchanged by or with the Committee that is related to a Committee recommendation to the MOP is to be made available to any Party on request.[39] In practice, the Ozone Secretariat has gone beyond these requirements in the non-compliance procedure by providing public access to Implementation Committee meeting reports, as well as the MOP reports that contain official compliance decisions, on the Internet.[40] Another way in which the Ozone Secretariat exceeds the requirements of decisions of the Parties is through efforts to make the primary meeting documents available in the UN official language preferred by each member of the Implementation Committee. Interpretation services in these languages are also offered.[41] A further valuable step taken to enhance transparency in the Montreal compliance system

34 J Wettestad, 'Monitoring and Verification' in D Bodansky, J Brunnée and E Hey (eds), *The Oxford Handbook of International Environmental Law* (Oxford University Press, 2007) 974, 979.

35 UNEP, 'Rules of Procedure for Meetings of the Conference of the Parties to the Vienna Convention and Meetings of the Parties to the Montreal Protocol', *Handbook for the Vienna Convention for the Protection of the Ozone Layer* (7th ed, 2006), rule 29 <http://ozone.unep.org/Publications/VC_Handbook/Section_3_Rules_of_Procedure/index.shtml>.

36 Non-compliance procedure, above n 26, para 15.

37 M Donaldson and B Kingsbury, 'The Adoption of Transparency Policies in Global Governance Institutions: Justifications, Effects, and Implications' (2013) 9 *Annual Review of Law and Social Science* 119, 120.

38 This was confirmed recently by the 'Our Common Vision' document, which was a product of the Rio+20 Conference, 'The Future We Want': Rio+20 Conference, 'Our Common Vision' (2012) para 77 <http://www.un.org/disabilities/documents/rio20_outcome_document_complete.pdf>; see also D B Hunter, 'The Emerging Norm of Transparency in International Environmental Governance' in P Ala'i and R G Vaughn (eds), *Research Handbook on Transparency* (Edward Elgar, 2014) 343.

39 Non-compliance procedure, above n 26, para 16.

40 Ozone Secretariat, *Handbook for the Montreal Protocol on Substances that Deplete the Ozone Layer,* above n 11.

41 Ozone Secretariat, *Primer for Members,* above n 9, 20.

is the production of the publicly available *Primer for Members* in 2007, the purpose of which is to outline both the formal non-compliance processes as well as the 'customary practices' of the Implementation Committee.[42] Hence, despite an ongoing commitment to safeguarding confidentiality, the Montreal compliance system has employed a range of procedures and practices to enhance transparency.

2.2.2 The implications of poor-quality data in national reports

Perhaps the most significant obstacle to the realization of internal transparency ideals in practice is chronic issues with the quality and reliability of states' reported information. The expectation that the Implementation Committee will respect states' reported information has contributed to these issues. Under paragraph 7(d) of the non-compliance procedure, the Implementation Committee has an obligation to 'identify the facts and possible causes relating to individual cases of non-compliance referred to the Committee'. In meeting this requirement, the Committee may request information from the Party concerned,[43] the Secretariat,[44] the Multilateral Fund or the Global Environment Facility, but it is expected that deference will be shown to information submitted by the Parties as sovereign states. This is exemplified by the findings of the Ad Hoc Working Group convened to review the non-compliance procedure in 1998:

> The Working Group also considered a proposal on the acquisition of additional data for verification purposes from other sources, including non-governmental organizations or private individuals. It was felt that the issue was already sufficiently covered in paragraph 3 of the non-compliance procedure ... Some experts pointed out that the Committee should seek information primarily from the concerned party and other formal sources, such as the Ozone Secretariat, lest conflicting information from different informal sources might make it difficult for the Implementation Committee to make recommendations. There was even a possibility that such relevant recommendations by the Committee might be rejected by the concerned party if based on information which that party found unacceptable.[45]

A similar stance is reflected in a statement made in the report of the President of the Implementation Committee at the seventh MOP in 1997: 'Regarding potential discrepancies between data submitted by a Party and data available elsewhere, ... the Committee had also agreed, as the matter concerned sovereign

42 *Ibid* 4.
43 With the Party's consent, on-site inspections may also be conducted: Non-compliance procedure, above n 26, para 7(e).
44 *Ibid* para 7(c).
45 UNEP, Report on the Work of the Ad Hoc Working Group of Legal and Technical Experts on Non-Compliance with the Montreal Protocol, above n 27, para 30.

States, that data submitted and supported by an individual country might be respected.'[46]

While the Implementation Committee's stance is respectful of state sovereignty, it effectively forecloses opportunities for independently verifying states' submitted data, drawing into question the accuracy and reliability of this information. Victor notes that a fundamental issue affecting the Montreal compliance system has been 'poor data quality because the system was never designed with the capacity to obtain independent data nor to confront countries that might be lying'.[47] States' willful misreporting of data is not unprecedented in MEAs, as the former Soviet Union's 'massive and systematic falsification of reports to the International Whaling Commission' demonstrates.[48] Despite these significant limitations, states' provision of imperfect data satisfies their procedural obligations for data reporting within the Montreal compliance system, and is used as a basis to evaluate their fulfillment of their substantive obligations to phase-out the production and consumption of ozone-depleting substances. In this regard, compliance decision making is strongly shaped by traditional diplomatic norms, at the expense of procedures of an administrative character that enhance transparency and accountability.

A related issue is poor-quality baseline data, which may be subsequently revisable. According to Article 7 of the Montreal Protocol, Parties are required to provide statistical 'baseline' data to the Secretariat on production, imports and exports for each of the controlled substances. It is against this baseline data that changes in Parties' production and consumption of ozone-depleting substances can be quantitatively assessed, underscoring its vital importance for subsequent compliance decisions. When Parties have been called to account to the Implementation Committee for potential non-compliance with their consumption and production targets under the Montreal Protocol, a common response has been that the yardstick against which their performance is being assessed is itself incorrect.[49] That is, the fault is claimed to lie with the inaccurately estimated baseline data, which is too low and thus should be revised upwards, rather than the Parties' non-compliant conduct.[50]

46 UNEP, Report of the President of the Implementation Committee on the Report of the Secretariat on Information Provided by the Parties in Accordance with Articles 4, 7 and 9 of the Montreal Protocol and on the Functioning of the Implementation Committee, UNEP/OzL.Pro.7/12, 27 Dec 1995, para 38.

47 Victor, *Global Warming Gridlock,* above n 33, 230. Similarly, Oberthür describes the data quality of states' reports as 'moderate at best': S Oberthür, *Production and Consumption of Ozone-Depleting Substances 1986-1999: The Data Reporting System under the Montreal Protocol* (Deutsche Gesellschaft für Technische Zusammenarbeit, 2001) 11–32.

48 A Chayes and A Handler Chayes, *The New Sovereignty: Compliance with International Regulatory Agreements* (Harvard University Press, 1998) 155.

49 Klabbers, 'Compliance Procedures', above n 17, 996.

50 *Ibid* 996 n 3.

It is likely that the requirements for classification as a developing country created an incentive for reporting low baseline data.[51] Article 5(1) specifies that, in order to be classified as a developing country, and therefore eligible for the benefits of delayed implementation and financial and technological transfer, a Party must be both a 'developing country' and have consumption of ozone-depleting substances below certain per capita thresholds. However, developing countries' performance was then measured against these 'unrealistically low' yardsticks, and unsurprisingly found wanting, leading to frequent requests for upwards revisions of baseline data.[52] Of course, a risk associated with such revisions is that they provide an easy way out for Parties that are in, or heading toward, non-compliance with their obligations.[53]

These types of requests have not been limited to developing countries and the early operation of the Montreal compliance system. Indeed, such requests are so common that Decision XV/19 of the fifteenth MOP in 2003 sets out a standardized methodology for reviewing Parties' requests to change their baseline data.[54] Implementation Committee and MOP reports indicate that assessing requests for revisions of baseline data are an ongoing and routine aspect of compliance work in the Montreal compliance system.[55] For example, Decision XXVI/14 adopted at the twenty-sixth MOP in 2014 approved revisions of baseline consumption data for HCFCs for Libya and Mozambique.[56] Therefore, issues with poor-quality data, which is not required to be independently verified, and

51 Bafundo claims that 'weak states' notoriety for corruption' creates 'a high likelihood of intentionally inaccurate data': N E Bafundo, 'Compliance with the Ozone Treaty: Weak States and the Principle of Common but Differentiated Responsibility' (2005–2006) 21 *American University International Law Review* 461, 486. Any such tendency is likely to be exacerbated by the absence of independent verification processes within the Montreal compliance system, discussed earlier in this sub-section.

52 Klabbers, 'Compliance Procedures', above n 17, 996 n 3.

53 D Brack, 'Monitoring the Montreal Protocol' (2003) *Verification Yearbook* 209, 219.

54 UNEP, Decision XV/19: Methodology for Submission of Requests for Revision of Baseline Data, UNEP/OzL.Pro.15/9, 11 Nov 2003.

55 See, eg, UNEP, Decision XIII/15: Data and Information Provided by the Parties to the 13th Meeting of the Parties in Accordance with Article 7 of the Montreal Protocol, UNEP/OzL.Pro.13/10, 26 Oct 2001; UNEP, Decision XIV/27: Request for Changes in Baseline Data, UNEP/OzL.Pro.14/9, 5 Dec 2002; UNEP, Decision XV/19: Methodology for Submission of Requests for Revision of Baseline Data, UNEP/OzL.Pro.15/9, 11 Nov 2003; UNEP, Decision XVI/31: Requests for Changes in Baseline Data, UNEP/OzL.Pro.16/17, 16 Nov 2004; UNEP, Decision XVIII/29: Request for Change in Baseline Data by Mexico, UNEP/OzL.Pro.18/10, 16 Nov 2006; UNEP, Decision XIX/24: Request for Change in Baseline Data by Turkmenistan, UNEP/OzL.Pro.19/7, 21 Sept 2007; UNEP, Decision XXIV/16: Requests for the Revision of Baseline Data by Algeria, Ecuador, Equatorial Guinea, Eritrea, Haiti, the Niger, the former Yugoslav Republic of Macedonia and Turkey, UNEP/OzL.Pro.24/10, 22 Nov 2012.

56 UNEP, Decision XXVI/14: Requests for the Revision of Baseline Data by Libya and Mozambique, UNEP/OzL.Pro.26/10, 10 Dec 2014.

may be subsequently revised, reflect transparency problems and draw into question the reliability of subsequent compliance decisions predicated on this data. In this regard, the procedural safeguards in place are not effectively constraining compliance processes that are subject to negotiation, undermining desirable aims of proceduralization.

2.3 Participation

The roles of experts, non-governmental organizations (NGOs) and civil-society representatives in the Montreal compliance system are limited. The key actors that are enabled to participate in the various stages of non-compliance processes are: the Implementation Committee members, the Ozone Secretariat, the Multilateral Fund Secretariat, the international and bilateral implementing agencies[57] of the Multilateral Fund and the Global Environment Facility, and the Party whose non-compliance matter is being heard before the Committee.[58] This sub-section argues that a primary purpose of the involvement of each of the actors in this process is to better understand the circumstances leading to a Party's non-compliance, and support its return to compliance. That is, the circumscribed participation roles in the Montreal compliance system promote regard for countries' individual circumstances, but are not intended to provide public accountability through extensive reliance on expert and civil-society inputs.

2.3.1 Non-decisional participation rights

In non-compliance proceedings, the Implementation Committee has decisional participation rights and the Party concerned has juridical participation rights. Non-decisional participation rights are afforded to representatives of the Ozone Secretariat, the Multilateral Fund, the Global Environment Facility and the Technology and Economic Assessment Panel. Since 1992, it has become common practice for the Multilateral Fund Secretariat and the Chair and Vice-Chair of the Executive Committee of the Multilateral Fund to be invited to Implementation Committee meetings as observers.[59] Moreover, when it became clear in the mid-1990s that CEITs were experiencing difficulties complying with their obligations under the Protocol, representatives of the Secretariat of the Global Environment Facility and the Technology and Economic Assessment Panel were invited

57 These are the United Nations Development Programme (UNDP), the United Nations Environment Programme (UNEP), the United Nations Industrial Development Organization (UNIDO) and the World Bank.

58 Ozone Secretariat, *Primer for Members,* above n 9, 8.

59 Multilateral Fund representatives first participated in an Implementation Committee meeting in September 1992: see UNEP, Report of the Fourth Meeting of the Implementation Committee, UNEP/OzL.Pro/ImpCom/4/2, 6 Oct 1992, para 2.

observers at Implementation Committee meetings.[60] The latter provided a degree of expert input into compliance decision making regarding options for addressing CEITs' non-compliance.[61]

As well as providing evidence to the Implementation Committee on compliance prospects and funding arrangements, the representatives of the Multilateral Fund and the Global Environment Facility often support the Parties involved in non-compliance proceedings. These representatives are typically invited by the Committee to present information on a developing country's or CEIT's future prospects for achieving compliance, and any past and planned financial assistance.[62] In addition, in practice these agencies often assist Parties to prepare documentation requested by the Committee, including compliance action plans,[63] and to implement measures recommended by the Committee to support the Party's return to compliance.[64] These representatives' dual role as advisers to both the Implementation Committee and the Party concerned underscores the supportive, amicable and facilitative nature of the non-compliance process.

There is no mechanism for members of the public to participate in Implementation Committee meetings. During the development of the non-compliance procedure, the possibility of the Implementation Committee considering information from NGOs, private entities and individuals was debated, but met with objections and was not incorporated into the procedure.[65] Previous requests by environmental NGOs to attend meetings of the Implementation Committee were denied on the basis that 'confidential, delicate, and sensitive information might be discussed and the presence of an NGO could limit frank discussion'.[66] There is, however, scope for NGOs and other non-state actors to participate by: making submissions on compliance to the Secretariat, lobbying the Executive Committee of the Multilateral Fund and the Implementation Committee to make Multilateral Fund assistance

60 At the thirty-ninth meeting of the Implementation Committee, the members expressed the view that the presence of a Global Environment Facility representative was desirable whenever the potential non-compliance of a Party receiving financial assistance from the Facility was being considered: UNEP, Report of the Thirty-Ninth Meeting of the Implementation Committee, UNEP/OzL.Pro/ImpCom/39/7, 11 Nov 2007, para 2.

61 UNEP, *Assessment of the Basic Problems Confronting Countries with Economies in Transition in Complying with the Montreal Protocol: Report of the TEAP Ad-Hoc Working Group on CEIT Aspects* (UNEP, 1995).

62 Ozone Secretariat, *Primer for Members*, above n 9, 10–11.

63 Brack, 'Monitoring the Montreal Protocol', above n 53, 223.

64 Ozone Secretariat, *Primer for Members*, above n 9, 11.

65 UNEP, Report of the Third Meeting of the Ad-Hoc Working Group of Legal and Technical Experts on Non-Compliance with the Montreal Protocol, UNEP/OzL.Pro/WG.3/3/3, 9 Nov 1991, para 23; Bankobeza, *Ozone Protection*, above n 4, 244.

66 Victor, 'The Operation and Effectiveness of the Montreal Protocol's Non-Compliance Procedure', above n 10, 142; see also J Tallberg, 'Orchestrating Enforcement: International Organizations Mobilizing Compliance Constituencies' in K W Abbott, P Genschel, D Snidal and B Zangl (eds), *International Organizations as Orchestrators* (Cambridge University Press, 2015) 166, 180.

contingent on compliance with specified benchmarks[67] and participating as observers in MOP meetings upon the President's invitation unless one-third of the Parties object.[68]

The Implementation Committee's approach to public participation is arguably unduly restrictive, particularly in the light of the more expansive public participation rules in other MEA compliance committees where sensitive information is discussed.[69] Handl asserts that one of the reasons the Implementation Committee resists NGO participation is that it could lead to 'an undue politicization of the committee's proceedings'.[70] An alternative view is that participation by experts and representatives of civil society can play an important role in providing a degree of 'representativeness' and giving voice to otherwise marginalized perspectives and interests in state-centric international decision making.[71] Regrettably, the latter was not an aim pursued by the institutional designers of the Montreal compliance system. The limited non-decisional participation rights afforded to civil-society actors significantly restrict opportunities for these actors to act as a check on power politics and facilitate procedural regard.

2.3.2 Juridical participation rights

Parties that are subject to the non-compliance procedure have juridical participation rights, which enhance regard for affected states' interests. Once the Secretariat becomes aware of a potential instance of non-compliance, it must notify the Party concerned within two weeks, and give that Party three months – or longer as may be required – to respond to the alleged facts.[72] If no reply is received within the three-month period, the Secretariat is to issue a reminder notification.[73] Once a

67 Bankobeza, *Ozone Protection*, above n 4, 232.
68 UNEP, Rules of Procedure, above n 35, rule 7. For example, at the twelfth MOP, attention was drawn to comments made by an 'observer from an environmental non-governmental organization' on the issue of phase-out of HCFCs for developing countries: UNEP, Report of the Twelfth Meeting of the Parties to the Montreal Protocol, UNEP/OzL.Pro.12/9, 10 Jan 2001, para 26.
69 C Pitea, 'NGOs in Non-Compliance Mechanisms under Multilateral Environmental Agreements: From Tolerance to Recognition' in T Treves et al (eds), *Civil Society, International Courts and Compliance Bodies* (TMC Asser Press, 2005) 204, 208–209; A Epiney, 'The Role of NGOs in the Process of Ensuring Compliance with MEAs' in U Beyerlin, P-T Stoll and W Rudiger (eds), *Ensuring Compliance with Multilateral Environmental Agreements: A Dialogue Between Practitioners and Academia* (Brill Martinus Nijhoff Publishers, 2006) 319, 320. See also the discussion of NGO participation in CITES meetings in Chapter 6, section 2.3.
70 G Handl, 'Compliance Control Mechanisms and International Environmental Obligations' (1997) 5 *Tulane Journal of International and Comparative Law* 29, 40–41.
71 See, eg, K Raustiala, 'NGOs in International Treaty-Making' in D B Hollis (ed), *The Oxford Guide to Treaties* (Oxford University Press, 2011) 150, 169.
72 Non-compliance procedure, above n 26, para 2.
73 *Ibid.*

matter is escalated for consideration by the Implementation Committee, the Party concerned may submit written information regarding its alleged non-compliance.[74] These procedures for notification and an opportunity to respond are consistent with administrative decision-making ideals.

In addition, Parties that are subject to non-compliance proceedings are given a right to be heard, which is 'one of the classical elements' of administrative law.[75] Where the non-compliance matter is complex or is likely to result in a compliance action plan, the Party is usually invited to participate in the Committee's consideration of the submission regarding its alleged non-compliance,[76] but not in the elaboration and adoption of the Committee's recommendations.[77] In practice, a key purpose of a Party representative attending a Committee meeting is to address any 'gaps' in the Committee's understanding of the Party's situation, and a special session is set aside during the meeting for consultation with the Party's representative for this purpose.[78] In other words, provision is made for individual review of non-compliance that is alert to contextual factors. These participation procedures thus facilitate procedural regard for the country-specific interests of affected states.

2.3.3 Decisional participation rights

The previous sub-section argued that non-decisional and juridical participation procedures pertaining to Implementation Committee meetings have been designed in a way that promotes regard for states' individual circumstances. In relation to both Implementation Committee and MOP meetings, *decisional* participation rights are vested in political representatives, which has not noticeably attenuated procedural regard for affected states' interests.

The members of the Implementation Committee serve as state representatives, which are selected in a way that promotes equitable representation of developed and developing countries' interests. During the development of the non-compliance procedure, there was extensive debate about whether members should serve as experts in their personal capacities. This proposal was ultimately unsuccessful as 'Parties were concerned that too much independence of individual members may cause the Meeting of the Parties to lose [control over the Committee] and the Meeting's decision-making power over matters of compliance'.[79] The ideal of 'equitable geographical distribution' guides the MOP's election of the ten Parties that sit on the Implementation Committee.[80] This means that there are two

74 Ozone Secretariat, *Primer for Members,* above n 9, 11.

75 B Kingsbury, N Krisch and R B Stewart, 'The Emergence of Global Administrative Law' (2005) 68 *Law and Contemporary Problems* 15, 37.

76 Non-compliance procedure, above n 26, para 10.

77 *Ibid* para 11.

78 Ozone Secretariat, *Primer for Members,* above n 9, 11, 26.

79 Bankobeza, *Ozone Protection,* above n 4, 261.

80 Non-compliance procedure, above n 26, para 5.

members from each of the five UN regional groups (Western Europe and others; Africa; Asia and the Pacific; Latin America and the Caribbean; and Eastern Europe), with one member from each group being replaced each year. The Committee also selects a President and a Vice-President, with Article 5 and non-Article 5 members alternating in these roles.[81] Therefore, although it is state representatives that vote on decisions in the Implementation Committee,[82] steps have been taken to minimize skews along North–South fault lines.

While the Implementation Committee is vested with the authority to make recommendations to the Parties on compliance issues, which have proved highly influential in practice, it is the MOP that is empowered to make final decisions on compliance matters.[83] In general, a majority of Parties present and voting is required to carry a vote.[84] However, in relation to the Russian Federation's non-compliance matter discussed at the MOP in 1995, a 'consensus minus one' voting rule was adopted. In this instance, the Secretariat clarified that, if only one Party objected to a draft decision, 'that decision would be carried by consensus and the position of the dissenting Party would be clearly reflected in the report of the Meeting'.[85] The vesting of ultimate oversight authority for decisions affecting the Parties in the MOP is consistent with practices in many MEAs,[86] which reflects the diminishing influence of traditional consent-based approaches in international environmental law.[87] Although the decision makers in the Implementation Committee and MOP are political representatives, the decision-making processes at the Implementation Committee stage in particular are procedurally constrained, and promote regard for affected states' interests.

2.4 Reason giving

As with the other administrative procedures discussed above, the reason-giving requirements in the Montreal compliance system allow and encourage the possible

81 Ozone Secretariat, *Primer for Members,* above n 9, 7–8.

82 In the absence of a set of procedural rules for the Implementation Committee, the Rules of Procedure of the MOP apply *mutatis mutandis*, including the requirement that decisions will be taken by consensus, or if consensus cannot be achieved, by a two-thirds majority vote of the Parties present and voting: UNEP, Rules of Procedure, above n 35, Rules 26(6) and 40(1).

83 Non-compliance procedure, above n 26, para 9.

84 The MOP Rules of Procedure require that decisions will be taken by consensus, or if consensus cannot be achieved, by a two-thirds majority vote of the Parties present and voting: UNEP, Rules of Procedure, above n 35, rule 40(1).

85 UNEP, Report of the Seventh Meeting of the Parties to the Montreal Protocol on Substances that Deplete the Ozone Layer, UNEP/OzL.Pro.7/12, 27 Dec 1995, para 130.

86 Cardesa-Salzmann, 'Constitutionalising Secondary Rules in Global Environmental Regimes', above n 9, 105.

87 See, eg, J Brunnée, 'COPing with Consent: Law-Making Under Multilateral Environment Agreements' (2002) 15(1) *Leiden Journal of International Law* 1; R R Churchill and G Ulfstein, 'Autonomous Institutional Arrangements in Multilateral Environmental Agreements: A Little-Noticed Phenomenon in International Law' (2000) 94 *American Journal of International Law* 623, 623.

causes of a Party's non-compliance to be acknowledged and contextualized. This sub-section argues that these requirements facilitate procedural regard for affected states' compliance interests. Moreover, similarly to the accountability procedures analyzed above, the practice of giving reasons has served to limit arbitrariness and promote consistency in compliance decision making, in line with administrative decision-making ideals. Hence, reason-giving procedures in the Montreal compliance system serve to constrain political influences and promote procedural regard.

When the Secretariat becomes aware of a possible instance of non-compliance, significant effort is made to understand and report upon its causes. These reports contain the Secretariat's reasons for escalating a non-compliance matter, and lay the groundwork for the Implementation Committee's subsequent consideration of the non-compliant Party's interests. The Secretariat is required to report to the Implementation Committee, orally and in writing, on the apparent deviation of the Party concerned, any response the Secretariat has received from that Party providing an explanation of the deviation and any other information regarding the Party's situation that the Secretariat deems useful for the Committee's deliberations.[88] The latter types of 'other information' may include, inter alia, the nature and status of assistance received from the Multilateral Fund or Global Environment Facility, and any extraneous circumstances, such as civil unrest or natural disasters, that are affecting the Party's capacity to comply.[89] The Secretariat's reason giving, supplemented by other relevant information from the Party concerned, and the Multilateral Fund or the Global Environment Facility, form the basis of the Implementation Committee's deliberations. These processes aim to ensure that the 'individual circumstances of each Party subject to the non-compliance procedure are taken into full consideration'.[90] Thus, extensive reason giving in the Secretariat's non-compliance referral process facilitates the Implementation Committee's regard for the country-specific interests of non-compliant Parties, which are frequently developing countries and CEITs.[91]

According to Stewart, the process of giving reasons is a powerful mechanism for disciplining political decision making, and evidencing procedural regard for weaker actors' interests.[92] The reason-giving practices in the Implementation Committee exhibit both of these qualities. In relation to the former, as noted above, there are sixteen standardized templates that the Committee uses for providing reasons for matters of routine procedural non-compliance. The use of the same or similar language for common contraventions promotes the equitable treatment of Parties in comparable circumstances,[93] which is consistent with administrative law

88 Ozone Secretariat, *Primer for Members,* above n 9, 14.
89 *Ibid.*
90 *Ibid* 31, 44.
91 Cardesa-Salzmann, 'Constitutionalising Secondary Rules in Global Environmental Regimes', above n 9, 115; Klabbers, 'Compliance Procedures', above n 17, 996–997.
92 Stewart, 'Remedying Disregard in Global Regulatory Governance', above n 5, 265.
93 Ozone Secretariat, *Primer for Members,* above n 9, 16, 45–58.

ideals of consistency and predictability.[94] However, the Committee also 'tailors the text of the recommendations to reflect the individual circumstances of the Parties concerned',[95] indicating that there is leeway for each state's unique situation to be factored into decision making, and for such considerations to be reflected in the published reasons for compliance decisions.

The decision on the non-compliance of France[96] in 2013 exemplifies both the individually tailored and formulaic elements of reason giving in the Montreal compliance system. The individualized elements of this decision relate to: (1) France's commitments under the Protocol, (2) the specific factual situations pertaining to France's non-compliance regarding excess production of HCFCs and (3) the submission of an action plan by France, which satisfied the MOP that France would return to compliance in subsequent years and that no further action was required.[97] This was followed by a number of formulaic statements: (1) that France's progress towards meeting the actions specified in its action plan will be 'closely monitored', (2) that it should 'continue to be treated in the same manner as a Party in good standing' as it works towards returning to compliance and (3) a caution that more serious consequences, including suspensions of rights and privileges under the Montreal Protocol, will be considered if France fails to return to compliance in a timely manner.[98] This combination of country-specific and standardized language simultaneously conveys procedural regard and consistency.

The fact that non-compliance recommendations were issued against Israel in 2016,[99] France in 2013[100] and the European Union (EU) in 2011,[101] is notable as it indicates that, at least in more recent times, the compliance process is not politicized to the extent that powerful developed countries escape accountability for non-compliance altogether.[102] Nonetheless, these remain atypical examples as developed country Parties have only infrequently been subject to non-compliance proceedings in the Montreal compliance system, with the vast majority of cases

94 D C Esty, 'Good Governance at the Supranational Level: Globalizing Administrative Law' (2006) 115 *Yale Law Journal* 1490, 1529.

95 Ozone Secretariat, *Primer for Members,* above n 9, 16.

96 UNEP, Decision XXV/11: Non-Compliance with the Montreal Protocol by France, UNEP/OzL.Pro.25/9, 13 Nov 2013.

97 *Ibid.*

98 *Ibid.* This is in accordance with item B of the Indicative List of Measures: see UNEP, Decision IV/5, above n 26, Annex V.

99 UNEP, Decision XXVIII/10: Non-compliance by Israel with its Data and Information Reporting Obligations, UNEP/OzL.Pro.28/12, 15 Nov 2016.

100 UNEP, Decision XXV/11: Non-compliance with the Montreal Protocol by France, above n 96.

101 UNEP, Decision XXIII/26: Non-compliance with the Montreal Protocol by the European Union, UNEP/OzL.Pro.23/11, 8 Dec 2011.

102 Cf the apparent favourable treatment of powerful states in the CITES compliance system: see Chapter 6, section 2.1.2.

concerning developing countries and, less frequently, CEITs.[103] As noted above, for these countries, recommendations are typically made for the non-compliant Party to continue to receive bespoke financial and technical assistance if it meets time-bound benchmarks specified in its action plan en route to full compliance. The more recent non-compliance proceedings involving developed countries suggest that power politics is not an overriding factor shaping decision-making processes in the Montreal compliance system.

2.5 Review

There is little scope for review built into the Montreal compliance system, which necessarily limits the role of review mechanisms in contributing to the desirable aims of proceduralization. The possibilities for review that do exist within this compliance system have not gained traction in practice. During the first review of the non-compliance procedure in 1998, the idea of appeals against decisions of the MOP on compliance matters was canvassed.[104] Although this proposal garnered little support, some experts opined that the proposal was simply premature, and should not be dismissed out of hand.[105] The issue has not been revisited, and as it currently stands, Parties have little recourse to review the compliance decisions made within the Montreal compliance system.

A second dimension of review relates to oversight of a compliance body's decisions by a superior body within the compliance hierarchy, which may occur in the absence of a formal appeal.[106] Although in theory the decisions of the Implementation Committee are subject to oversight by the MOP, in practice the MOP's review function is largely symbolic, as it routinely endorses the Committee's compliance recommendations.[107] Despite this paucity of actively utilized review mechanisms, the preceding discussion has accentuated the positive roles played by other administrative procedures in promoting regard for the compliance interests of affected states.

3 A vignette of compliance decision making: the Russian Federation

The non-compliance proceedings concerning the Russian Federation provide an apt vignette to illustrate the practical operation of the Montreal compliance system. These proceedings exemplify compliance decision making that reflects some constraints on political decision making, and procedural regard. They are

103 See Cardesa-Salzmann, 'Constitutionalising Secondary Rules in Global Environmental Regimes', above n 9, 115; Klabbers, 'Compliance Procedures', above n 17, 996–997.

104 UNEP, Report on the Work of the Ad Hoc Working Group of Legal and Technical Experts on Non-compliance with the Montreal Protocol, above n 27, para 53(c).

105 *Ibid* para 57.

106 See Chapter 2, section 3.5.

107 Ozone Secretariat, *Primer for Members,* above n 9, 18.

one of the few instances in which enforcement measures have been utilized. This represents a departure from a predominantly facilitative approach, and arguably constitutes an example of politicized decision making that is unusual in the compliance system's history. This vignette also highlights the Implementation Committee and MOP's willingness to allow interest representation and to extend conditional financial assistance to CEITs, which reflects procedural and substantive regard for these countries' compliance interests. This created a precedent for future non-compliance responses for CEITs, entrenching these states' expectations of procedural regard buttressed by financial and technical assistance.

Concerns about the Russian Federation's compliance with the Montreal Protocol were evident a number of years before formal non-compliance proceedings commenced in 1995. Russia's compliance difficulties were already apparent in 1992 when it announced that it was 'experiencing extraordinary political, economic and social difficulties and did not have the capacity to assume additional obligations under the new amendments and adjustments to the Montreal Protocol'.[108] Russia argued that it would direct all available resources toward meeting targets for ozone-depleting substances in its industry, and thus would be unable to contribute to the Multilateral Fund. Moreover, it requested 'a specific preferential regime in the Protocol's implementation', or else it would object to further restrictions on ozone-depleting substances under the Protocol.[109]

Despite these warning signs, Russia's case was not reviewed in detail until its non-compliance became the subject of proceedings before the Implementation Committee in 1995.[110] At this time, Russia was on the verge of non-compliance, and in a statement made on behalf of itself and four other Parties (Belarus, Bulgaria, Poland and Ukraine) and three future Parties (Armenia, Georgia and Kyrgyzstan), Russia requested a five-year grace period to comply with its treaty obligations.[111] This statement was originally addressed to the MOP, which re-directed it to the Implementation Committee, underscoring the latter body's pivotal role in compliance decision making. The Implementation Committee interpreted Russia's submission to the MOP as self-reporting on non-compliance, triggering paragraph 4 of the non-compliance procedure.[112]

108 UNEP, Report of the Fourth Meeting of the Parties, UNEP/OzL.Pro.4/15, 25 Nov 1992, para 82.

109 *Ibid.*

110 J Werksman, 'Compliance and Transition: Russia's Non-Compliance Tests the Ozone Regime' (1996) 56 *Zeitschrift für Ausländisches Recht und Völkerrecht* 750, 764.

111 UNEP, Report of the Implementation Committee under the Non-Compliance Procedure for the Montreal Protocol on the Work of its Eleventh Meeting, UNEP/OzL.Pro/ImpCom/11/1, 14 Sept 1995, Annex II.

112 F Romanin Jacur, 'The Non-Compliance Procedure of the 1987 Montreal Protocol to the 1985 Vienna Convention on Substances that Deplete the Ozone Layer' in T Treves et al (eds), *Non-Compliance Procedures and Mechanisms and the Effectiveness of International Environmental Agreements* (TMC Asser Press, 2009) 11, 15, 31.

The Implementation Committee's response was consistent with the above-mentioned emphasis on understanding the particular interests and capacity constraints affecting individual countries. To better understand the implementation challenges faced by the Parties listed in Russia's statement, the Implementation Committee sought expert advice from a special Technology and Economic Assessment Panel on countries with economies in transition.[113] Thereafter, the Implementation Committee concluded a compliance action plan with each Party containing, inter alia, a statement by the Party of its political commitment to achieve compliance, identification of the steps necessary to achieve compliance, and an agreement on what assistance was required for the Party to implement these agreed steps.[114] Hence, considerable effort was invested in understanding the country-specific factors contributing to non-compliance, and thereby affording procedural regard.

The agreed compliance action plans and the Implementation Committee's recommendations were subsequently considered, and largely followed, by the MOP. In relation to Russia, the Implementation Committee recommended restrictions on exports of ozone-depleting substances, to which the Russian Federation expressed strong reservations.[115] Adopting a consensus-minus-one voting procedure, the MOP restricted Russia's trade in such substances to members within the Commonwealth of Independent States,[116] rather than imposing a blanket ban on exports. Russia deemed these 'discriminatory measures and sanctions' to be 'unacceptable' and arbitrary, as there had been no prior recourse to the facilitative options specified in the Indicative List of Measures.[117] Indeed, as previously mentioned, it is arguable that these trade restriction measures were not authorized by this list at all.[118] Furthermore, as Werksman observes, other objections that Russia may have raised include:

> [T]hat the invocation of the Non-Compliance Procedure's Indicative List of Measures was premature, as Russia was not ... formally found in non-compliance; that the scope of the ban, which refers generally to substances controlled by the Montreal Protocol, would appear to include substances added in the Copenhagen Amendments, to which Russia [was] not yet a

113 Victor, 'The Operation and Effectiveness of the Montreal Protocol's Non-Compliance Procedure', above n 10, 157.

114 Gillespie, *Climate Change, Ozone Depletion, and Air Pollution,* above n 29, 222.

115 UNEP, Report of the Seventh Meeting of the Parties, above n 85, para 44.

116 UNEP, Decision VII/18: Compliance with the Montreal Protocol by Russia, UNEP/OzL. Pro.7/12, 27 Dec 1995, para 8.

117 UNEP, Report of the Seventh Meeting of the Parties, above n 85, paras 128–129.

118 M Fitzmaurice, 'Non-Compliance Procedures and the Law of Treaties' in T Treves et al (eds), *Non-Compliance Procedures and Mechanisms and the Effectiveness of International Environmental Agreements* (Cambridge University Press, 2009) 453, 473. For a discussion of the implications of Article 60 of the Vienna Convention on the Law of Treaties in this context, see *ibid* 472–474.

> signatory; and that by requiring Russia to 'secure that no exports will be made from the [Commonwealth of Independent States]' the Meeting of the Parties was asking that Russia be held responsible for activities beyond its territorial control.[119]

Thus, it appears that the trade restrictions imposed on Russia, as well as Belarus[120] and Ukraine[121] at the same MOP, were an anomaly in an otherwise consistently facilitative approach to non-compliance in the Montreal compliance system. This departure from procedural guidelines in the non-compliance procedure and past practices helps to explain the Russian Federation's indignation[122] about the apparent arbitrariness and politicized nature of the recommended measures.[123]

In addition to these intrusive measures, the MOP offered bespoke financial and technical assistance to support Russia's return to compliance.[124] This is consistent with the earlier precedents set in responding to developing countries' non-compliance, evidencing an administrative style of decision making. The disbursement of international financial assistance to Russia was made contingent upon the settlement of reporting and other compliance issues with the Implementation Committee.[125] Russia was held to account for the funds received against tiered benchmarks negotiated with the Implementation Committee, creating a precedent that was followed for other CEITs in non-compliance.[126] As previously noted, this proceduralized approach for taking into account the specific compliance interests of CEITs promotes procedural and substantive regard. The availability of financial and technical assistance valuably reinforces these procedural regard opportunities.

This form of flexible accountability does, however, come at the cost of a Party's direct accountability for its binding treaty commitments, underscoring the difference between proceduralization and legalization.[127] The Russian Federation

119 Werksman, 'Compliance and Transition', above n 110, 768–769.

120 UNEP, Decision VII/17: Compliance with the Montreal Protocol by Belarus, UNEP/OzL.Pro.7/12, 27 Dec 1995, para 6.

121 UNEP, Decision VII/19: Compliance with the Montreal Protocol by Ukraine, UNEP/OzL.Pro.7/12, 27 Dec 1995, para 6.

122 Brack notes that the Russian delegation walked out of the 1995 MOP where these issues were being discussed: Brack, 'Monitoring the Montreal Protocol', above n 53, 219.

123 The perception that this was a case of brazen, serious and deliberate violation of the Montreal Protocol may help to explain the MOP's departure from procedural guidelines and past practices in this instance: see, eg, J Peel, 'New State Responsibility Rules and Compliance with Multilateral Environmental Obligations: Some Case Studies of How the New Rules Might Apply in the International Environmental Context' (2001) 20(1) *Review of European Community and International Environmental Law* 82, 94.

124 UNEP, Decision VII/18, above n 116, para 9; Werksman, 'Compliance and Transition', above n 110, 769–771.

125 UNEP, Decision VII/18, above n 116, para 9(e).

126 See above n 28.

127 In the context of domestic regulatory systems, Dubash and Morgan have similarly observed that proceduralization does not necessarily equate to legalism: N Dubash and B Morgan,

remained in non-compliance with various provisions of the Montreal Protocol from 1996 to 2000, and was also in non-compliance with the benchmarks specified in its own compliance action plan in 1999 and 2000.[128] It was not until 2002 that the MOP confirmed that the Russian Federation had returned to compliance with its obligations.[129] Nevertheless, from 1995 to 2002, Russia continued to be treated as a 'country in good standing' and to receive financial assistance from the Global Environment Facility.[130] This is an example of how compliance arrangements may accord with desirable aims of proceduralization while simultaneously accommodating circumvention of the rules of international law regarding breach of treaty.[131]

4 Conclusion

This chapter has provided evidence of the adoption of a circumscribed set of administrative procedures for accountability, transparency, participation and reason giving in the Montreal compliance system. The specific administrative procedures employed are well suited to promoting procedural regard for affected states' interests, including those of less powerful states. The design of the compliance apparatus reflects an emphasis on understanding the reasons for Parties' non-compliance, and supporting their return to compliance, through flexible yet procedurally standardized approaches. In line with this approach, some administrative procedures that are effective for constraining political influences on decision making, such as independent verification processes for states' reported information, expert decision making, participation by civil-society representatives, and appeal mechanisms, have not been incorporated in the Montreal compliance system. One undesirable consequence of these gaps is that the consistency and reliability of processes within the Montreal compliance system, which are often premised on poor-quality data, are drawn into question.

Nonetheless, over time, a shift towards an administrative style of decision making, evidenced through a consistent and formulaic approach to addressing

'The Embedded Regulatory State: Between Rules and Deals' in N Dubash and B Morgan (eds), *The Rise of the Regulatory State of the South: Infrastructure and Development in Emerging Economies* (Oxford University Press, 2013) 279, 292.

128 UNEP, Decision XIII/17: Compliance with the Montreal Protocol by the Russian Federation, UNEP/OzL.Pro.13/10, 26 Oct 2001, para 2.

129 UNEP, Report of the Fourteenth Meeting of the Parties to the Montreal Protocol on Substances that Deplete the Ozone Layer, UNEP/OzL.Pro.14/9, 5 Dec 2002, para 133.

130 See UNEP, Decision VIII/25: Compliance with the Montreal Protocol by the Russian Federation, UNEP/OzL.Pro.8/12, 19 Dec 1996; UNEP, Decision IX/31: Compliance with the Montreal Protocol by the Russian Federation, UNEP/OzL.Pro.9/12, 25 Sept 1997; UNEP, Decision X/26: Compliance with the Montreal Protocol by the Russian Federation, UNEP/OzL.Pro.10/9, 3 Dec 1998; UNEP, Decision XIII/17: Compliance with the Montreal Protocol by the Russian Federation, UNEP/OzL.Pro.13/10, 26 Oct 2001.

131 For further analysis of this issue, see Chapter 7, section 2.3; and Huggins, 'Administrative Procedures and Rule of Law Values in the Montreal Compliance System', above n 21.

non-compliance for developing countries and CEITs, is discernible. Attempts to balance the aspiration of individualized responses to non-compliance with the aim of similar treatment for comparable cases reflect a mode of administrative decision making that preserves some discretionary leeway for decision makers. In particular, the accountability arrangements for conditional financial assistance from the Multilateral Fund and the Global Environment Facility have contributed to standardized responses to substantive compliance issues for developing countries and CEITs, which are reflected in a formulaic approach to reason giving in compliance decisions. These conditional funding arrangements reinforce procedural regard opportunities by providing states with the financial and technical means to address their compliance issues and interests that have been explored through the non-compliance process.

5 The Kyoto compliance system and the potential for procedurally constrained decision making*

1 Introduction

The Kyoto compliance system is remarkable for its attempts to constrain political influences,[1] and create an autonomous, arm's length and rules-based compliance process. This compliance system incorporates a number of components: (1) requirements for national measurement and reporting of Annex I states' emissions inventories, (2) internationally co-ordinated external verification and review of national emissions inventories by expert review teams (ERTs), (3) the resolution of compliance issues and determination of consequences for non-compliance by the regime's Compliance Committee, which consists of the Bureau, plenary and Facilitative and Enforcement Branches and (4) ultimate oversight by the Conference of the Parties serving as the Meeting of the Parties (CMP).[2] The Kyoto compliance system thus provides a sophisticated administrative apparatus for review of state action by independent bodies,[3] highlighting its suitability as a case study for evaluating the benefits of proceduralization.

The institutional design of the Kyoto compliance system reflects attempts to insulate the compliance hierarchy from undue political influence. In particular, the Facilitative and Enforcement Branches were designed to undertake compliance determinations in a procedurally constrained environment. The Facilitative Branch has been tasked with advising on and facilitating implementation for all Parties taking into account 'the circumstances pertaining to the questions before it', and

* Parts of this Chapter build upon analysis in A Huggins, 'The Desirability of Depoliticization: Compliance in the International Climate Regime' (2015) 4(1) *Transnational Environmental Law* 101.

1 G Ulfstein, 'Depoliticizing Compliance' in J Brunnée, M Doelle and L Rajamani (eds), *Promoting Compliance in an Evolving Climate Regime* (Cambridge University Press, 2012) 418, 418.

2 J Brunnée, 'Climate Change and Compliance and Enforcement Processes' in R Rayfuse and S V Scott (eds), *International Law in the Era of Climate Change* (Edward Elgar, 2012) 290, 303–304; A Zahar, 'Verifying Greenhouse Gas Emissions of Annex I Parties: Methods We Have and Methods We Want' (2010) 1 *Climate Law* 409, 411.

3 A Zahar, J Peel and L Godden, *Australian Climate Law in Global Context* (Cambridge University Press, 2012) 126.

providing an early warning function in relation to Annex I states' potential non-compliance.[4] In contrast, the Enforcement Branch ensures states' accountability with their commitments under the Kyoto Protocol through automatic and non-discretionary responses corresponding with different types of compliance issues. Taken together, these two Branches provide an institutional framework that includes options for regard for affected states' interests and strong mechanisms to hold states to account for non-compliance.

In practice, however, the role of the Facilitative Branch has been effectively bypassed as facilitative roles have been assumed by ERTs,[5] which has created a bottleneck limiting the number of cases heard before the Enforcement Branch. The degree to which ERTs assumed these facilitative roles was perhaps unanticipated by the institutional designers of the Kyoto compliance system, who created separate Facilitative and Enforcement Branches with wide-ranging powers that were safeguarded by numerous due-process guarantees.[6] In contrast, ERTs' processes have been relatively opaque and unclear, raising questions about whether the system as implemented has been adequately responsive to affected states' interests. This analysis highlights the desirability of utilizing procedurally constrained decision-making spaces where they are available to prevent a shift to more informal negotiating spaces that lack due-process guarantees. A vignette of the non-compliance proceedings involving Greece in section 3 reinforces the dissonance between this compliance system's extensive proceduralization and actual decision-making practices, which has been exacerbated by technical experts in the ERTs assuming facilitative roles for which they are insufficiently equipped.

While most of the analysis in this chapter applies to both the Kyoto Protocol's first (2008–2012) and second (2013–2020) commitment periods, the primary focus is on the first commitment period that included a greater number of state participants, and in which the Kyoto Protocol remained a lynchpin of the international climate-change regime. Some of the rules in the compliance system have changed during the course of the second commitment period;[7] where this has occurred, the primary focus is on the lessons that can be learned from the

4 Decision 27/CMP.1, Procedures and Mechanisms Relating to Compliance under the Kyoto Protocol, FCCC/KP/CMP/2005/8/Add.3, 30 Mar 2006, section IV, paras 4 and 6.

5 A Zahar, *International Climate Change Law and State Compliance* (Routledge, 2015) Chapter 3.

6 Rather, as Yoshida notes, it was hoped that ERTs would play a role in depoliticizing compliance: O Yoshida, 'Procedural Aspects of the International Legal Regime for Climate Change: Early Operation of the Kyoto Protocol's Compliance System' (2011) 4 *Journal of East Asia and International Law* 41, 46.

7 For example, the guidelines for expert review under the Kyoto Protocol applied in the first commitment period (see, eg, Decision 22/CMP.1, Guidelines for Review under Article 8 of the Kyoto Protocol, FCCC/KP/CMP/2005/8/Add.3, 30 Mar 2006) were updated for application in 2016: see, eg, Decision 4/CMP.11, Revisions to the Guidelines for Review under Article 8 of the Kyoto Protocol, FCCC/KP/CMP/2015/8/Add.1, 29 Jan 2016, Annex I. See also the discussion of International Assessment and Review for Annex I countries in Chapter 7, section 3.

experience of implementing the relevant rules that applied during the first commitment period. The central role played by the Kyoto Protocol has diminished since the successful negotiation of the Paris Agreement in 2015, leading some to argue that, despite its continued operation, the Kyoto Protocol's second commitment period has already run its course.[8] Although the Kyoto Protocol is unlikely to continue operation after the end of the second commitment period, this analysis provides transferable insights for the development of the Paris Agreement's compliance apparatus.[9]

2 Evaluating proceduralization in the Kyoto compliance system

The analysis in this section applies the proceduralization framework to the Kyoto compliance system. Accountability and participation mechanisms feature prominently in the design of this compliance system, and accordingly are foregrounded in the following discussion.

2.1 Accountability in the Kyoto compliance system

2.1.1 ERTs and deference to state sovereignty in review practices

ERTs are the first tier of oversight for states' performance information in the compliance hierarchy. If an ERT identifies that a state has failed to comply with its emissions obligations under the Kyoto Protocol, they are required to refer the compliance issue for further consideration by the Compliance Committee. Prima facie, therefore, the elements of accountability – an accounter required to justify conduct, an account holder to whom the accounter is answerable and consequences for conduct that does not meet specified standards[10] – are all present in the Kyoto compliance system's review processes. In practice, however, ERTs' role as a regulatory intermediary between the Compliance Committee and individual Parties has gone well beyond strict monitoring and verification, and has encompassed elements of dialogue, rule interpretation and negotiations.[11] This sub-section argues that the procedures for impartial and autonomous technical review processes by ERTs have been undermined by simultaneous expectations on ERTs to respect state sovereignty and the diplomatic customs of international law. This has had

8 See, eg, A Zahar, 'A Bottom-Up Compliance Mechanism for the Paris Agreement', *Chinese Journal of Environmental Law* (forthcoming, 2017) 8 <https://papers.ssrn.com/sol3/papers.cfm?abstract_id=2912560>.

9 See Chapter 7, section 3.

10 R B Stewart, 'Remedying Disregard in Global Regulatory Governance: Accountability, Participation, and Responsiveness' (2014) 108 *American Journal of International Law* 211, 253.

11 On the topic of the roles played by regulatory intermediaries generally, see K W Abbott, D Levi-Faur and D Snidal, 'Theorizing Regulatory Intermediaries: The RIT Model' (2017) 670 *Annals of the American Academy of Political and Social Science* 14, 22.

deleterious consequences for the reliability and consistency of ERT review processes, and restricted the number of cases heard before the procedurally constrained Compliance Committee.

The review requirements to be fulfilled by ERTs are articulated in Article 8 of the Protocol. This includes a requirement at sub-paragraph (3) that '[t]he review process shall provide a thorough and comprehensive technical assessment of all aspects of the implementation by a Party of [the] Protocol'. In the first commitment period, inventories were to be reviewed against the basic principles of transparency, consistency, comparability, completeness and accuracy, as required by the United Nations Framework Convention on Climate Change (UNFCCC) reporting guidelines.[12] The main focus of such review processes, which consisted primarily of desk reviews, and may have included in-country reviews, was the overall reliability of national emissions inventories.[13]

Despite the ostensibly technical nature of expert review of states' national emissions inventories, considerable deference to state sovereignty has been expected in the review process. In practice, there is considerable uncertainty associated with quantifying greenhouse gas (GHG) emissions, leading to reliance on estimation techniques,[14] which can undermine the accuracy and completeness of states' emissions reports. This means that while reviewed emissions inventories are deemed to be legally compliant, they may not be 'in scientific compliance'.[15] The main checks that ERTs have been able to make are: (1) comparisons with a state's historically reported data, (2) conformity with standard Intergovernmental Panel on Climate Change methodologies, (3) country-level statistics on the production, import and export of fuel from the International Energy Agency and (4) comparisons with the types of issues reported in other states' reports.[16] Non-state sources are not to be consulted in verifying states' emissions information unless the government of the state under review formally supplies that data.[17] In other words, ERTs have been able to compare the consistency of information that

12 UNFCCC Secretariat, Review of the Implementation of Commitments and of Other Provisions of the Convention: UNFCCC Guidelines on Reporting and Review, FCCC/CP/2002/8, 28 Mar 2003, Section B, para 2.

13 A Zahar, 'Does Self-Interest Skew State Reporting of Greenhouse Gas Emissions? A Preliminary Analysis Based on the First Verified Emissions Estimates under the Kyoto Protocol' (2010) 1 *Climate Law* 313, 315.

14 G Ulfstein and J Werksman, 'The Kyoto Compliance System: Towards Hard Enforcement' in O S Stokke, J Hovi and G Ulfstein (eds), *Implementing the Climate Regime: International Compliance* (Earthscan, 2005) 39, 52.

15 T Berntsen, J Fuglestvedt and F Stordal, 'Reporting and Verification of Emissions and Removals of Greenhouse Gases' in O S Stokke, J Hovi and G Ulfstein (eds), *Implementing the Climate Regime: International Compliance* (Earthscan, 2005) 85, 85.

16 In the first commitment period, four reports were scrutinized by an ERT in a six-day, centralized review process: Zahar, Peel and Godden, *Australian Climate Law in Global Context,* above n 3, 104–106.

17 UNFCCC Secretariat, *Handbook for Review of National GHG Inventories* (undated), 11–12, cited in Zahar, Peel and Godden, *Australian Climate Law in Global Context*, above n 3, 119.

states have reported to various international bodies, and undertake procedural and other types of consistency checks. These powers fall short of robust, independent verification of the emissions information reported by states.

ERTs' reliance on official materials produced by sovereign states contrasts with the more wide-ranging powers for collecting compliance information that are available to the Enforcement Branch.[18] These restricted sources of information have placed a significant limitation upon ERTs' capacity to meaningfully review the accuracy and completeness of states' emissions reports. Considering the significant authority that ERTs have assumed within the Kyoto compliance system detailed below, it is concerning that their ability to provide rigorous review of states' emissions information has been so constrained. As the following analysis demonstrates, this is just one of a number of ways in which deference to state sovereignty has been expected of ERTs.

2.1.2 The reliability and consistency of expert review processes

The CMP has expected that ERT review processes will reflect a high degree of political neutrality and rigor. This expectation is codified in the following requirement of Decision 22/CMP.1:

> Each expert review team shall provide a thorough and comprehensive technical assessment of information submitted under Article 7 and shall, under its collective responsibility, prepare a review report, assessing the implementation of the commitments of the Party included in Annex I and identifying any potential problems in, and factors influencing, the fulfilment of commitments. The expert review teams *shall refrain from making any political judgement*.[19]

However, guidelines issued by the UNFCCC in 2003 state that one of the aims of ERT review should be the examination of reported information in a 'facilitative and open manner'.[20] As diplomacy and negotiations are almost invariably intertwined with facilitative processes,[21] it is contradictory to expect ERTs to engage in compliance facilitation while refraining from 'making any political judgement'. There is therefore a tension between the UNFCCC's 2003 guidance and the CMP's decision in 2006.

Although it may be expected that the latter decision would override the former guidance, the UNFCCC's guidance appears to have significantly influenced

18 See Chapter 5, section 2.1.4.
19 Decision 22/CMP.1, Guidelines for Review under Article 8 of the Kyoto Protocol, above n 7, para 21 (emphasis added).
20 UNFCCC Secretariat, 'Review of the Implementation of Commitments', above n 12, 83.
21 J Klabbers, 'Compliance Procedures' in D Bodansky, J Brunnée and E Hey (eds), *The Oxford Handbook of International Environmental Law* (Oxford University Press, 2007) 995, 1001, 1007–1009.

practice. Zahar, Peel and Godden observe that the state retains 'much of its sovereign power and the UNFCCC Secretariat has carefully managed the ERTs to ensure that their attitude is facilitative and respectful of the age-old customs of international law'.[22] This has included 'considerable give and take' between the ERT and the state under review during a period of facilitative dialogue in which the state can voluntarily revise its emission accounts to align with preliminary advice received from ERTs.[23] States' ability to *revise* reported data in line with ERT advice seems to go well beyond one of the stated objectives of the ERT process, which is to 'assist' Parties to *improve* their emissions reporting and the implementation of their commitments under the Protocol.[24]

In addition to states' ability to revise their national reports, ERTs have been able to revise their review reports in the light of their facilitative dialogue with states. Fransen's interview data collected from ERT members conducting reviews of Annex I national communications under the UNFCCC suggests that 'parties at times pressure the review teams to alter the language used in the [ERT] reports'.[25] As the same pool of reviewers[26] is responsible for reviews of Annex I states' national emissions inventories, and the compliance stakes are higher, this tendency is also likely to be evident, and even more pronounced, in the Kyoto compliance system. That is, despite the procedural rule stating that ERTs shall refrain from making any political judgement, diplomatic and facilitative decision-making modes have considerable traction in practice.

The way that ERTs exercise discretion in escalating compliance matters has also raised consistency concerns. There is a clear procedure for ERTs' listing of questions of implementation for referral to the Compliance Committee:

> Only if an unresolved problem pertaining to language of a mandatory nature in these guidelines influencing the fulfillment of commitments still exists after the Party included in Annex I has been provided with opportunities to correct the problem within the time frames established under the relevant review procedures, *shall* that problem be listed as a question of implementation in the final review reports.[27]

22 Zahar, Peel and Godden, *Australian Climate Law in Global Context*, above n 3, 104.
23 *Ibid*; Decision 27/CMP.1, above n 4, para 7.
24 Decision 27/CMP.1, above n 4, para 2(c) (emphasis added).
25 T Fransen, 'Enhancing Today's MRV Framework to Meet Tomorrow's Needs: The Role of National Communications and Inventories' (World Resources Institute, 2009) 8 <www.wri.org/sites/default/files/national_communications_mrv.pdf>.
26 ERTs are responsible for 'in-depth review' of Annex I Parties' national communications, and technical review of their inventories, under the UNFCCC. They are also responsible for the 'periodic' reviews of national communications and the 'annual reviews' of inventories under the Kyoto compliance system: Zahar, *International Climate Change Law and State Compliance*, above n 5, 41.
27 Decision 27/CMP.1, above n 4, para 8 (emphasis added).

This requirement could be interpreted to mean that ERTs must list non-compliance with a mandatory requirement of the Kyoto Protocol as a question of implementation if the issue is not resolved through dialogue.[28] This interpretation aligns with accountability aims to the extent that it promotes the consistent treatment of non-compliance issues in accordance with pre-agreed standards.

However, in practice, the rules of procedure regarding ERTs' listing of questions of implementation for referral to the Compliance Committee have been construed as vesting discretion in the ERTs to determine *if* and *when* they will list a Party's breach of a mandatory rule for action by the Committee.[29] Significantly, ERTs have assumed an ad hoc gatekeepers' role, referring only eight matters to the Compliance Committee in the first commitment period, and one to date in the second commitment period. These referrals seem highly unlikely to reflect the full range of compliance issues that have arisen. One explanation for this is that ERTs have been engaging in facilitation themselves,[30] rather than passing such matters up to be dealt with by the Facilitative Branch. A related explanation is that states are more willing to negotiate with ERTs to resolve differences than to have the matter escalated so as to require a formal compliance determination.[31] In either case, the net result is that the procedurally constrained compliance processes of the Compliance Committee have been bypassed in favor of more opaque ERT processes that lack extensive due-process guarantees.

A risk associated with these informal facilitative processes is that consistency in addressing compliance issues is undermined. There have been multiple attempts by the Compliance Committee to address this issue. For example, in the 2011 *Annual Report* of the Compliance Committee, the plenary 'recommended' that future ERT reports include a list of problems identified in the review process, providing reasons as to whether or not each problem relates to language of a mandatory nature, and if the ERT decides not to list such a problem as a question of implementation, an explanation of the basis for this decision.[32] This signaled that the plenary was seeking to enhance ERTs' consistency and accountability through imposing reason-giving requirements. Further, in 2012, both Branches and the Bureau of the Compliance Committee proposed a joint workshop with ERT lead reviewers focused on the issue of improving the consistency of

28 Zahar, 'Verifying Greenhouse Gas Emissions of Annex I Parties', above n 2, 420. The requirements for facilitative dialogue are provided in Decision 27/CMP.1, above n 4, para 7.

29 Zahar, 'Verifying Greenhouse Gas Emissions of Annex I Parties', above n 2, 422.

30 M Doelle, 'Early Experience with the Kyoto Compliance System: Possible Lessons for MEA Compliance System Design' (2010) 1(2) *Climate Law* 237, 260.

31 R Lefeber and S Oberthur, 'Key Features of the Kyoto Protocol's Compliance System' in J Brunnée, M Doelle and L Rajamani (eds), *Promoting Compliance in an Evolving Climate Regime* (Cambridge University Press, 2012) 77, 94.

32 Compliance Committee, Annual Report of the Compliance Committee to the Conference of the Parties Serving as the Meeting of the Parties to the Kyoto Protocol, Seventh Session, FCCC/KP/CMP/2011/5, 3 Nov 2011, para 28.

reviews,[33] which was held in Bonn, Germany, in March 2013.[34] A follow-up meeting was held in March 2015.[35] Thus, there has been significant concern about the appropriateness of ERTs' decision-making processes, and concerted attempts to rectify this issue.

2.1.3 The bypassing of the Facilitative Branch

In contrast to the ERTs, the Facilitative Branch provides a forum that is procedurally constrained, and designed to take affected states' interests into account. This Branch is tasked with advising on and facilitating implementation for all Parties, taking into account the principle of common but differentiated responsibilities and respective capacities.[36] In addition, it is required to 'take into account the circumstances pertaining to the questions before it', reinforcing that states' capacities and circumstances are mandatory considerations in the decision-making process. The Facilitative Branch also has an early warning function for Annex I states' potential non-compliance with their obligations under the Protocol, and can provide advice and facilitation for states at risk of failing to meet their quantified mitigation targets.[37] As is explained below, there are numerous procedural safeguards built into the design of the Facilitative Branch, all of which are prima facie conducive to constraining politics and enhancing regard.

However, the Facilitative Branch's potential in this regard has effectively remained latent. In the first commitment period, the Bureau allocated all eight substantive compliance matters referred by the ERTs to the Enforcement Branch. Only one substantive submission was made to the Facilitative Branch by South Africa on behalf of the G-77 and China in 2006 – that is, before the start of the first commitment period. This submission did not proceed due to the Branch members' failure to agree on a procedural issue.[38] This means that the Facilitative Branch's potential role in holding states to account for their obligations under the Kyoto Protocol has remained untested.[39]

33 Compliance Committee, Annual Report of the Compliance Committee to the Conference of the Parties Serving as the Meeting of the Parties to the Kyoto Protocol, Eighth Session, FCCC/KP/CMP/2012/6, 8 Nov 2012, para 28.

34 Compliance Committee, Annual Report of the Compliance Committee to the Conference of the Parties Serving as the Meeting of the Parties to the Kyoto Protocol, Ninth Session, FCCC/KP/CMP/2013/3, 1 Oct 2013, para 7.

35 Compliance Committee, Annual Report of the Compliance Committee to the Conference of the Parties Serving as the Meeting of the Parties to the Kyoto Protocol, Eleventh Session, FCCC/KP/CMP/2015/3, 18 Sept 2015, para 8.

36 Decision 27/CMP.1, above n 4, section IV, para 4.

37 *Ibid* para 6.

38 M Doelle, 'Compliance and Enforcement in the Climate Change Regime', in E J Hollo, K Kulovesi and M Mehling (eds), *Climate Change and the Law* (Springer, 2012) 165, 171–172.

39 For analysis of the role of the Facilitative Branch since the first commitment period ended in 2012, see Zahar, 'A Bottom-Up Compliance Mechanism for the Paris Agreement', above n 8, 12–15.

The limited recourse to the Facilitative Branch is partly attributable to the pivotal role played by ERTs in facilitating compliance,[40] which has rendered the role of the Facilitative Branch largely obsolete. The Secretariat, which as previously noted, carefully manages ERTs 'to ensure their attitude is facilitative and respectful of the age-old customs of international law',[41] appeared to endorse the assumption of facilitative responsibilities by ERTs in the Facilitative Branch's place. An additional factor contributing to the Facilitative Branch's moribund state in the first commitment period was the lack of a trigger for its early warning function.[42] Since 2011, the Branch has attempted to rectify this issue and clarify its mandate by re-interpreting the Branch's rules in a way that gives it an effective role in advice provision and addressing early warning issues. Despite this, it has achieved minimal success in establishing a meaningful niche for itself vis-à-vis the facilitative roles played by the ERTs.[43] Thus, the Facilitative Branch's role in holding Annex I states to account for their obligations has been negligible.

The assumption of facilitative power by the ERTs in the place of the Facilitative Branch has resulted in reduced opportunities for procedurally constrained decision making. As is elaborated below, although ERTs must take into account information submitted by the Parties, and provide some conclusions and reasons in their ERT reports, there are far fewer procedural safeguards governing ERTs' work compared to the Facilitative Branch.[44] Despite the extensive proceduralization in the Compliance Committee, a relative lack of procedural rigor at the ERT gatekeeper stage has undermined the potential benefits of proceduralization in the compliance system as a whole.

2.1.4 The Enforcement Branch and accountability

The Enforcement Branch has recourse to a significantly wider array of performance information sources than the ERTs. These sources include information provided by the Party concerned and ERTs, the Conference of the Parties (COP) to the UNFCCC, the CMP, the subsidiary bodies, competent intergovernmental organizations and non-governmental organizations (NGOs) and external experts.[45] Therefore, the capacity of the Enforcement Branch to meaningfully review Parties' compliance with their obligations under the Protocol is considerably greater than that of the ERTs. This is appropriate to legitimate the authority to impose intrusive

40 Lefeber and Oberthur, 'Key Features of the Kyoto Protocol's Compliance System', above n 31, 99.

41 Zahar, Peel and Godden, *Australian Climate Law in Global Context*, above n 3, 104.

42 S Oberthur, 'Options for a Compliance Mechanism in a 2015 Climate Agreement' (2014) 4 *Climate Law* 30, 40.

43 Zahar, *International Climate Change Law and State Compliance*, above n 5, 72–74, 79–82.

44 See further Chapter 5, section 2.3.1.

45 Decision 27/CMP.1, above n 4, sections VIII.3 and VIII.4; Decision 4/CMP.2, Compliance Committee, FCCC/KP/CMP/2006/10/Add.1, 4 Mar 2007, rule 20.

consequences for non-compliance vested in the Enforcement Branch.[46] The practical utility of these extensive review powers is, however, drawn into question given the bottleneck created by the ERTs, whose decisions are based on far more limited information.

The design and practice of the Enforcement Branch reflect the accountability norm of like treatment of analogous cases, rather than discretionary leeway to take into account country-specific interests and circumstances. Unlike the flexible approaches that typically characterize MEA compliance systems,[47] the rules of the Enforcement Branch provide an 'automatic review approach' for different types of non-compliance.[48] As noted in Chapter 3, these rules include the requirement of deductions from future emissions allocations if a Party's emission target is exceeded, a 'compliance action plan' for remedying non-compliance with methodological and reporting requirements, and states' suspension from participating in the Protocol's flexibility mechanisms if the non-compliance issue concerns the eligibility requirements.[49] This consistent approach has been maintained despite specific requests by Croatia[50] and Ukraine[51] for a degree of flexibility to apply to countries undergoing the process of transition to a market economy. In both cases, the Enforcement Branch reasoned that it was not within its competence to grant flexibility in the absence of a decision from the CMP.[52]

Consistent consequences were applied in each of the eight substantive compliance matters decided before the Enforcement Branch in the first commitment period.[53] Six of the eight compliance matters involved eligibility requirements, simultaneously raising questions of methodology and reporting, meaning that the consequences applied included both exclusion from the Protocol's flexibility mechanisms and a requirement to produce a compliance action plan. In the case of Canada heard in 2008, the Enforcement Branch deemed that Canada had rectified the factual issues that originally catalyzed the question of implementation in relation to reporting and methodological requirements, and no consequences were applied.[54] In the

46 See Chapter 3, section 5.
47 See, for example, the discussion of the Montreal compliance system in Chapter 4.
48 Brunnée, 'Climate Change and Compliance and Enforcement Processes', above n 2, 306.
49 Decision 27/CMP.1, above n 4, section XV, paras 5(a), (b) and (c). See also Chapter 3, section 3.1.4.
50 Compliance Committee, Written Submission from Croatia, CC-2009-1-5/Croatia/EB, 9 Oct 2009.
51 Compliance Committee, Further Written Submission from Ukraine, CC-2011-2-8/Ukraine/EB, 28 Sept 2011.
52 Compliance Committee, Decision on Preliminary Examination (Party Concerned: Croatia), CC-2009-1-2/Croatia/EB, 8 Sept 2009; Compliance Committee, Final Decision (Party Concerned: Ukraine), CC-2011-2-9/Ukraine/EB, 12 Oct 2011.
53 The eight substantive matters that were heard before the Enforcement Branch in the first commitment period concerned Greece, Canada, Croatia, Bulgaria, Romania, Ukraine, Lithuania and Slovakia.
54 The inability of the Kyoto compliance system to address or resolve Canada's non-compliance with its first commitment period target ultimately led to Canada's withdrawal from the Kyoto

2012 case of Slovakia, which involved non-compliance with methodological and reporting requirements, the consequence applied was the requirement for a compliance action plan to be submitted within three months. Therefore, the consequences applied treat analogous cases similarly according to ascertainable rules, which is consistent with procedurally constrained decision-making and accountability ideals. Notably, however, this automatic approach does not reflect the responsiveness to contextual considerations that is characteristic of procedural regard.

2.2 Transparency in the Kyoto compliance system

The accountability issues pertaining to compliance facilitation analyzed above also have implications for transparency. The opaqueness and informality associated with the ERTs' processes would be reduced if compliance facilitation occurred in the more procedurally constrained Facilitative Branch. This sub-section focuses on additional issues relating to the internal and external transparency of compliance information, and the language and technical barriers that may be limiting procedural regard.

Vis-à-vis other global MEA compliance systems,[55] the Kyoto compliance system has a relatively strong record on transparency. Parties' reports containing national emissions information, required under Article 7 of the Kyoto Protocol, are made available on the UNFCCC website.[56] These reports are then reviewed by ERTs under Article 8, and the resultant review reports are also made available to other states and to the public via the UNFCCC website if they raise a question of implementation,[57] promoting both internal and external transparency. Similarly, all decisions of the Compliance Committee branches are made public, as is all information considered by the relevant branch, although the branch or Party concerned may request a delay in making certain information available until the proceedings have concluded.[58]

The most significant obstacles to the effectiveness of these transparency policies in practice relate to the volume and technicality of the information provided, and language barriers, all of which undermine transparency aims associated with proceduralization. As an example of the former, some national inventory reports

Protocol in 2011, and resulted in 'heavy criticism of the compliance system as a whole': Lefeber and Oberthur, 'Key Features of the Kyoto Protocol's Compliance System', above n 31, 99.

55 See, eg, Chapters 4 and 6.

56 UNFCCC, 'Initial Reports under Article 7, paragraph 4, of the Kyoto Protocol and Initial Review Reports' (2017) <http://unfccc.int/national_reports/initial_reports_under_the_kyoto_protocol/first_commitment_period_2008-2012/items/3765.php>.

57 UNFCCC, 'Compliance under the Kyoto Protocol' (2017) <http://unfccc.int/kyoto_protocol/compliance/items/2875.php>.

58 Lefeber and Oberthur, 'Key Features of the Kyoto Protocol's Compliance System', above n 31, 89.

have been in excess of 1000 pages in length,[59] and it is often the case that more accurate emissions inventory methods are highly technically complex and difficult to understand.[60] The technical information on questions of implementation that is publicly available on the UNFCCC website[61] can be similarly dense. The Rules of Procedure also specify that the working language of the Compliance Committee shall be English,[62] and final decisions of the branches are to be translated into all official languages of the United Nations (English, Arabic, Chinese, French, Russian and Spanish).[63] If the transparent compliance information provided is technically dense and/or in a foreign language, the capacity of some actors, including those from less powerful states, to meaningfully engage with this information will be limited. These transparency challenges hamper opportunities for procedural regard.

2.3 Participation in the Kyoto compliance system

Although the types of non-decisional participation rights catered for are limited,[64] the Kyoto compliance system has well-developed juridical and decisional participation rights. Both these latter types of participation rights are significant from a procedural regard perspective, and hence are a focus of the following analysis. This sub-section argues that the bypassing of the Facilitative Branch, and limited recourse to the Enforcement Branch's juridical participation rights and

59 A Herold, 'Experiences with Articles 5, 7, and 8: Defining the Monitoring, Reporting and Verification System under the Kyoto Protocol' in J Brunnée, M Doelle and L Rajamani (eds), *Promoting Compliance in an Evolving Climate Regime* (Cambridge University Press, 2012) 122, 127–128.

60 *Ibid.* Tiered methodological approaches to preparing emissions inventories are provided by specified guidance materials produced by the Intergovernmental Panel on Climate Change, the use of which was mandatory in the first commitment period (Article 5(2)): *ibid* 125.

61 UNFCCC, 'Compliance under the Kyoto Protocol', above n 57.

62 Decision 4/CMP.2, above n 45, rule 13.1. Meetings of the Committee that are held in public are also recorded and broadcast on the Internet through the UNFCCC website: UNFCCC, 'Webcasts and Videos: Compliance Committee' (2017) <https://unfccc.int/press/multimedia/webcasts/items/5862.php>.

63 Decision 4/CMP.2, above n 45, rule 13.3.

64 According to rule 9 of the Compliance Committee's Rules of Procedure, except for the 'elaboration and adoption of a decision of a branch', 'meetings of the plenary and the branches shall be held in public, unless the plenary or branch of its own accord or the request of the Party concerned decides, for overriding reasons, that part or all of the meeting shall be held in private': Decision 4/CMP.2, above n 45, rule 9. A subsequent Working Arrangements rule stipulates that 'observers will be required not to interfere in any way with the conduct of the meeting, not to make contact with members or alternate members of the Committee while the session is ongoing and to leave the room when the plenary or the facilitative or enforcement branch decides to hold the meeting in private', reinforcing the strictly non-decisional participation rights of external participants in relation to Compliance Committee proceedings: Compliance Committee, Annual Report of the Compliance Committee to the Conference of the Parties Serving as the Meeting of the Parties to the Kyoto Protocol, Third Session, FCCC/KP/CMP/2007/6, 26 Sept 2007, para 17.

procedural safeguards has curtailed opportunities for procedural regard. Furthermore, an important dimension of participation in the Kyoto compliance system is the extensive reliance on expert decision making. While this reflects a concerted effort to insulate compliance processes from undue political interference, a strong reliance on experts is potentially problematic when the roles that experts are expected to play have political and facilitative dimensions that go beyond experts' traditional skill set.

2.3.1 Juridical participation rights

Once a non-compliance issue has been referred to either the Facilitative Branch or the Enforcement Branch, there are a number of juridical participation rights afforded to the relevant Party. These include:[65] (1) making information available to the Party concerned and requiring notifications to be sent to the Party at the different stages of the process,[66] (2) allowing the Party the opportunity to comment in writing on all information considered by the relevant branch, as well as on any decisions,[67] (3) permitting the Party to designate persons to represent it during the consideration of the question of implementation by the relevant branch,[68] (4) allowing written submissions, including rebuttal, from the Party[69] and (5) providing the Party with a hearing, to be held in public unless otherwise decided, where it may present its views, and expert testimony or opinion.[70] There is also an expedited procedure for questions of implementation pertaining to the flexibility mechanisms heard before the Enforcement Branch.[71] The procedural rules applying to both branches represent a high level of due-process guarantees compared to other MEA compliance systems.[72]

65 This list is adapted from the list of procedural requirements in: UNFCCC Secretariat, Procedural Requirements and the Scope and Content of Applicable Law for the Consideration of Appeals under Decision 27/CMP.1 and Other Relevant Decisions of the Conference of the Parties Serving as the Meeting of the Parties to the Kyoto Protocol, as well as the Approach Taken by other Relevant International Bodies Relating to Denial of Due Process, Technical Paper, FCCC/TP/2011/6, 15 Sept 2011, 7.

66 Decision 27/CMP.1, above n 4, annex, section VII, paras 4 and 5; section VIII, para 7; and section IX, para 6.

67 *Ibid* annex, section VII, para 7; and section VIII, paras 6 and 8; and Decision 4/CMP.2, above n 45, rules 17 and 18.

68 Decision 27/CMP.1, above n 4, annex, section VIII, para 2; and Decision 4/CMP.2, above n 45, rule 25(3).

69 Decision 27/CMP.1, above n 4, annex, section IX, paras 1 and 7; and Decision 4/CMP.2, above n 45, rule 17.

70 Decision 27/CMP.1, above n 4, annex, section IX, para 2; and Decision 4/CMP.2, above n 45, rule 9(1).

71 Decision 24/CP.7, Procedures and Mechanisms Relating to Compliance under the Kyoto Protocol, FCCC/CP/2001/13/Add.3, 21 Jan 2002, annex, section X, para 1.

72 See, eg, M Montini, 'Procedural Guarantees in Non-Compliance Mechanisms' in T Treves et al (eds), *Non-Compliance Procedures and Mechanisms and the Effectiveness of International Environmental Agreements* (TMC Asser Press 2009) 389.

The design of the Facilitative and Enforcement Branches thus reflects extensive opportunities for juridical participation rights and procedural safeguards to contribute to fair proceedings and a right to be heard. However, the minimal recourse to the Facilitative and Enforcement Branches in practice means that these opportunities have not been fully realized, with deleterious consequences for procedural regard for affected states' interests. The Facilitative Branch is the only branch of the Compliance Committee that was designed to apply to developing countries, ensuring that differentiated substantive obligations were reinforced by the compliance system's structure.[73] Yet, as most developing-country obligations are currently reflected in the UNFCCC, rather than the Kyoto Protocol, the Facilitative Branch has had limited practical relevance for developing countries[74] – or indeed developed countries.[75] The Enforcement Branch, which covers Annex I states' commitments, only heard eight matters during the first commitment period, and one thereafter. Accordingly, the procedural-regard opportunities built into the design of the Compliance Committee have had limited utility in practice.

2.3.2 Decisional participation rights and expert review

ERTs have decisional participation rights within the Kyoto compliance system, yet there is a considerable disconnect between their technical roles evidenced in the compliance rules, and the realities of their facilitative roles and decision making in practice. The compliance rules and procedures require members of ERTs to serve in their personal capacities – that is, as technical experts, rather than as state representatives[76] – and they cannot participate in reviews of their country of origin.[77] To achieve geographical representation, ERTs are typically composed of six experts from diverse countries, excluding the country under review. The Annex I countries to which the first and second commitment periods have applied are all developed countries and CEITs. Nonetheless, in determining the composition of these teams, the Secretariat has aimed to ensure an appropriate balance between representatives from Annex I and non-Annex I Parties,[78] as well as a geographical balance within these two groups.[79] Taken together, these requirements prima facie promote ERT members' impartiality.

73 According to Rajamani, this is evident from 'the provisions relating to the mandate of the enforcement branch which cover Annex I commitments alone (i.e. compliance with Articles 3.1, 5.1 and 5.2, 7.1, and 7.4, and eligibility requirements under Articles 6, 12, and 17)': L Rajamani, 'Developing Countries and Compliance in the Climate Regime' in J Brunnée, M Doelle and L Rajamani (eds), *Promoting Compliance in an Evolving Climate Regime* (Cambridge University Press, 2012) 367, 391.

74 *Ibid* 388.

75 See Chapter 5, section 2.1.3.

76 Decision 22/CMP.1, above n 7, para 23.

77 *Ibid* para 25.

78 Of the two lead reviewers for each review, one is from an Annex I and one is from a non-Annex I Party.

79 Herold, 'Experiences with Articles 5, 7, and 8', above n 59, 135.

However, ERT members' political independence is drawn into question if they are effectively playing a double role as reviewers and those whose work is reviewed. The UNFCCC Roster of Experts[80] reveals that almost all reviewers work in government departments for their national governments, and many are involved in the preparation of their own country's emissions inventory.[81] Fransen's 'not-for-attribution interviews' with individuals involved in the review process for Annex I national communications under the UNFCCC[82] indicate that ERT members are hesitant to dispute each other's communications 'for fear of their own communications being challenged'.[83] Again, as the compliance stakes are higher under the Kyoto Protocol, which has binding emissions targets, this tendency is likely to be more pronounced in the Kyoto compliance system. The practices that allow reviewers to serve this double role are out of step with the spirit of the rules and procedures, which codify more rigorous standards of impartiality.

The appropriateness of technical experts engaging in diplomatic facilitation of compliance deserves further scrutiny. It is notable that the designers of the Kyoto compliance system departed from common practice in other major global MEAs, including the Montreal Protocol and CITES, where UN-sponsored bureaucracies or political bodies play key roles in facilitating compliance.[84] The assumption of facilitative power by the ERTs under the Kyoto compliance framework raises a question about whether the technical experts conducting expert reviews are in fact suitably equipped to be engaging in the high-level politics associated with facilitating states' compliance with their international environmental commitments. While most reviewers are national government bureaucrats, a minority have come from other career backgrounds such as academia.[85] Such backgrounds may equip members well to deal with the technical dimensions of emissions review. However, if ERTs have been encouraged by the Secretariat to play a pivotal role in facilitating compliance, it is arguable that a background in international diplomacy and negotiations is beneficial, and perhaps indispensable,[86] for engaging in high-stakes political compliance negotiations.

A more open acknowledgment of the dual nature of ERTs' functions, and the limits of technical expertise in these political settings, raises questions about

80 UNFCCC, 'UNFCCC Roster of Experts' (2017) <http://www4.unfccc.int/sites/roe/Pages/Home.aspx>.

81 Zahar, 'Verifying Greenhouse Gas Emissions of Annex I Parties', above n 2, 423–424.

82 See above nn 25 and 26.

83 Fransen, 'Enhancing Today's MRV Framework', above n 25, 8.

84 See, eg, A Fodella, 'Structural and Institutional Aspects of Non-Compliance Mechanisms' in T Treves et al (eds), *Non-Compliance Procedures and Mechanisms and the Effectiveness of International Environmental Agreements* (TMC Asser Press, 2009) 355, 360.

85 UNFCCC Secretariat, 'Review of the Implementation of Commitments', above n 12, 83.

86 On the types of knowledge, skills and attitudes that are valuable in international environmental diplomacy and negotiations, see S Bankobeza, E Maruma Mrema and K Ramakrishna, 'International Environmental Diplomacy and Negotiations' in T Kuokkanen, E Couzens, T Honkonen and M Lewis (eds), *International Environmental Law-making and Diplomacy: Insights and Overviews* (Routledge, 2016) 83, 88–95.

whether multidisciplinary teams, composed of both technical experts and skilled diplomats/negotiators, would be better equipped to fulfill ERTs' diverse responsibilities. Alternatively, ERTs could be limited to undertaking more technical and procedural roles, in line with their expertise. This shift would require clarification and re-specification of the rules and guidelines pertaining to ERTs to remove references to 'facilitative' roles.[87] Such changes would provide greater scope for the Facilitative Branch to assume its procedurally constrained facilitative role in accordance with its mandate, which aligns with desirable aims of proceduralization.

2.3.3 Decision makers and decision-making processes in the Compliance Committee

Like the ERTs, the composition of the Compliance Committee is designed to convey impartiality. Both the Facilitative and Enforcement Branches are composed of one member from each of the five UN regional groups – one nominated by a small island developing country, two nominated by Annex I developed countries and two nominated by non-Annex I developing countries.[88] This means that 60 per cent of the members of the Compliance Committee are nominated by developing countries,[89] despite the fact that only Annex I industrialized countries have binding emission reduction obligations.[90] In a further effort to ensure that compliance decisions are not strongly shaped by geopolitical considerations, a simple majority of votes among the developed and developing country members is required to pass a decision in the Enforcement Branch.[91] This is in addition to a requirement that the four functional formations of the Compliance Committee attempt to take decisions by consensus, with decisions by a three-quarters majority permitted as a last resort.[92] Hence, numerous procedural safeguards have been implemented to minimize the risk that decision making reflects political biases.

Moreover, there are requirements for the expert members of the Compliance Committee to be independent. Members of the Compliance Committee 'shall serve in their individual capacities'[93] as technical experts, rather than as state representatives, and shall 'act in an independent and impartial manner and avoid

87 See further Chapter 7, section 3.

88 Decision 27/CMP.1, above n 4, section V, para 1.

89 Lefeber and Oberthur, 'Key Features of the Kyoto Protocol's Compliance System', above n 31, 80.

90 Werksman notes that this was a contentious aspect of the negotiations regarding the development of the Kyoto compliance system: J Werksman, 'The Negotiation of a Kyoto Compliance System' in O Schram Stokke, J Hovi and G Ulfstein (eds), *Implementing the Climate Regime: International Compliance* (Earthscan, 2005) 17, 28.

91 Lefeber and Oberthur, 'Key Features of the Kyoto Protocol's Compliance System', above n 31, 80.

92 Decision 27/CMP.1, above n 4, section II, para 9.

93 *Ibid* section II, para 6.

real or apparent conflicts of interest'.[94] Furthermore, they must take a written oath of services, vouching that their roles will be undertaken 'honourably, faithfully, impartially and conscientiously' and with full disclosure of any potential conflicts of interest.[95] A complaint procedure has been established for alleged conflicts of interest or incompatibility with 'the requirements of independence and impartiality'.[96] These are desirable procedural safeguards for persons exercising decision-making authority over states.

These aspirations of impartiality may be undermined by the risks of partisanship stemming from domestic funding sources and national interest representation. In the case of developed states,[97] due to a lack of centrally provided funding, it is the Party who nominated the expert that pays for their expenses, and '[s]ome governments have questioned whether they should provide such reimbursement if they cannot instruct the member or alternate member nominated by them to serve the interests of that state'.[98] Repeated requests for the CMP to provide funding for all regular and alternate members of the Enforcement Branch in the interests of independence and neutrality have been unsuccessful.[99] A further risk of partisanship stems from the fact that members of the Compliance Committee can concurrently serve as a member of a delegation to a meeting under the UNFCCC or the Kyoto Protocol. This appeared to be an issue in an appeal made to the CMP that was initiated by Croatia in 2010, which was subsequently withdrawn with no reasons provided.[100] In the appeal documentation, it was claimed that a conflict of interest existed, as an alternate member of the Enforcement Branch 'was also a member of the EU delegation at COP-12 in Nairobi which had expressed its reservation regarding the applicability of the flexibility under decision 7/CP.12 for Croatia to the Kyoto Protocol'.[101] The Compliance Committee has recommended that 'due diligence' be exercised in such potential conflict of interest cases.[102] It is recognized that these types of partisanship risks may beset all international organizations.[103]

94 Decision 4/CMP.2, above n 45, rule 4.1.
95 *Ibid* rule 4.2.
96 *Ibid* rule 4.4.
97 The travel and subsistence expenses of representatives from some developing and low-income countries are reimbursed by the Secretariat: Lefeber and Oberthur, 'Key Features of the Kyoto Protocol's Compliance System', above n 31, 83–84.
98 *Ibid* 84.
99 *Ibid* 83–84.
100 Decision 14/CMP.7, Appeal by Croatia against a Final Decision of the Enforcement Branch of the Compliance Committee in Relation to the Implementation of Decision 7/CP.12, FCCC/KP/CMP/2011/10/Add.2, 15 Mar 2012, para 1.
101 Compliance Committee, Annual Report of the Compliance Committee to the Conference of the Parties Serving as the Meeting of the Parties to the Kyoto Protocol, Sixth Session, FCCC/KP/CMP/2010/6, 8 Oct 2010, para 53; see also Ulfstein, 'Depoliticizing Compliance', above n 1, 420.
102 Compliance Committee, Annual Report of the Compliance Committee, Sixth Session, above n 101, para 50.
103 Fodella, 'Structural and Institutional Aspects of Non-Compliance Mechanisms', above n 84, 362–363.

As the analysis in this chapter shows, such risks have not significantly impacted on the procedural rigor of the Compliance Committee's processes.

In sum, public participation in the Kyoto compliance system is limited to non-decisional participation, states and expert consultants are afforded juridical participation rights and technical and legal experts have decisional participation rights in the ERTs and the Compliance Committee. Attempts to create politically insulated zones for decision making have been significantly more successful in the Compliance Committee than in the ERTs, yet decision-making power has tended to concentrate with the latter.

2.4 Reason giving in the Kyoto compliance system

2.4.1 Reason giving by expert review teams

Relatively high levels of reason giving in the Kyoto compliance system are evident at both the ERT and Compliance Committee stages. While in the case of the ERTs, there has been a lack of consistency in the reasons provided, reason giving in the Enforcement Branch is largely consistent with administrative decision-making ideals.

ERTs are required to provide reasons for their review decisions. In the first commitment period, all final review reports prepared by ERTs were required to include the following elements: (1) an introduction and summary, (2) a description of the technical assessment of each of the elements reviewed in accordance with the *Guidelines for Review under Article 8 of the Kyoto Protocol*, including the identification of 'potential problems, [and] factors influencing the fulfillment of commitments', recommendations to solve any such problems, and an assessment of the Annex I Party's efforts to address these problems, (3) recommendations for the conduct of future reviews, (4) identification of other issues of concern and (5) the sources of information used in formulating the final report.[104] These reports thus incorporate relevant factual and legal considerations, and final recommendations.

The consistency of reason giving in practice has, however, been undermined by ERTs' facilitative processes. As previously noted, after the Party concerned has been provided with opportunities to correct the problem, ERTs are required to list 'an unresolved problem pertaining to language of a mandatory nature' in an ERT report.[105] In relation to this requirement, the Enforcement Branch formally expressed concern about the 'lack of clarity' in a 2010 ERT report regarding Bulgaria. In March 2010, an ERT finalized its review of Bulgaria's 2009 inventory report, and identified a question of implementation concerning Bulgaria's national system, which triggered non-compliance proceedings. Specifically, the ERT

104 Decision 22/CMP.1, above n 7, Part I, section F, para 48.
105 *Ibid* para 8.

concluded that Bulgaria's national system did not operate in accordance with the Guidelines for National Systems for the Estimation of Emissions by Sources and Removals by Sinks under Article 5(1) of the Kyoto Protocol. This finding was informed by inadequacies in: (1) the country's institutional arrangements and (2) the arrangements for the technical competence of staff within the national system involved in the inventory development process.[106]

Significantly, these problems were not new for Bulgaria, and had been identified in both in-country and desk reviews by ERTs in the previous two years.[107] In line with their mandate to assist states to improve their national emissions reporting,[108] in previous years the ERTs made suggestions for enhancing Bulgaria's national system, but only engaged the Compliance Committee in relation to these concerns in 2010. In its final decision concerning Bulgaria, the Enforcement Branch formally expressed concern about the 'lack of clarity' in this ERT report regarding why unresolved problems had not been previously listed as questions of implementation:

> During its deliberations, the branch noted with concern the lack of clarity in the 2010 [Annual Review Report], which does not clearly explain why unresolved questions did not result in the listing of questions of implementation pursuant to paragraph 8 of the annex to decision 22/CMP.1. In particular, differing interpretations of this provision may lead to different conclusions as to whether an unresolved problem is required to be listed as a question of implementation. This reveals more systematic issues that concern the review process under Article 8 of the Kyoto Protocol and the compliance system as a whole, which require urgent attention.[109]

This comment by the Enforcement Branch underscores the opaqueness of ERTs' facilitative roles, and implies a rebuke of ERTs' exercise of discretion in deciding if and when to list questions of implementation and thus escalate compliance matters. Hence, the consistency of reasons provided by the ERTs has been problematic, and has detracted from the realization of administrative decision-making ideals associated with reason giving.

2.4.2 Reason giving in the Compliance Committee

Similarly to the ERT reports, there are procedural requirements that all preliminary and final decisions of the Compliance Committee contain certain elements,

106 Compliance Committee, Report of the Review of the Annual Submission of Bulgaria Submitted in 2009, CC-2010-1-1/Bulgaria/EB, 17 Mar 2010, para 200.
107 Doelle, 'Compliance and Enforcement in the Climate Change Regime', above n 38, 182.
108 Decision 22/CMP.1, above n 7, paras 2(c), 5 and 7.
109 Compliance Committee, Decision under Paragraph 2 of Section X (Party Concerned: Bulgaria), CC-2010-1-17/Bulgaria/EB, 4 Feb 2011, para 14.

including conclusions and reasons for the decision.[110] After the initial hearing, the Enforcement Branch makes a preliminary finding as to whether the Party is in compliance, after which the Party has an opportunity to comment on this preliminary finding. If comments are filed from the Party through this process, the Enforcement Branch then issues a final decision, providing reasons for its decision in the light of the comments received.[111] The final decision must be communicated to the Party concerned in writing, and made available to other Parties and the public.[112]

The Enforcement Branch's reason-giving procedures and practices both reflect and reinforce the non-discretionary nature of compliance decision making. Rule 22 of the Rules of Procedure specifies the information that a preliminary finding or a final decision is to contain.[113] In the final decision issued for the compliance matters heard before the Enforcement Branch, there are three sections: (1) a background section, which is typically brief, (2) a conclusions and reasons section, which elaborates the rules under the Kyoto Protocol and secondary rules and procedures, as well as the factual considerations, that are relevant to the compliance determination and (3) a decision section, which communicates the final decision and its consequences.[114] The whole document is typically two to three pages in length. There is little evidence of dialogue regarding competing considerations discussed during the hearings.[115] This lack of justification is not unexpected given that the application of consequences for certain types of non-compliance is intended to be automatic.[116] This sense of automaticity is reinforced by the fact that reasons for dissenting votes may be provided in the report of the meeting, but not in the final decision.[117] Thus, reason-giving practices in the Enforcement Branch accord with some of the desirable aims of proceduralization, yet do not reflect the flexibility to take into account individual countries' circumstances in determining compliance consequences that typically inheres in procedural regard.

110 Decision 27/CMP.1, above n 4, section VIII, para 7; and section IX, paras 5 and 9; and Decision 4/CMP.2, above n 45, rules 12 and 22(1)(g).

111 Decision 27/CMP.1, above n 4, section IX, para 7; and section X, para 1(e); and Decision 4/CMP.2, above n 45, rule 17.

112 Decision 27/CMP.1, above n 4, section VIII, para 7.

113 Decision 4/CMP.2, above n 45, rule 22.

114 See, eg, Compliance Committee, Decision under Paragraph 2 of Section X (Party Concerned: Greece), CC-2007-1-13/Greece/EB, 13 Nov 2008; Compliance Committee, Decision under Paragraph 2 of Section X (Party Concerned: Croatia), CC-2009-1-14/Croatia/EB, 8 Feb 2012; Compliance Committee, Decision under Paragraph 2 of Section X (Party Concerned: Bulgaria), CC-2010-1-17/Bulgaria/EB, 4 Feb 2011.

115 M Doelle, 'Experience with the Facilitative and Enforcement Branches of the Kyoto Compliance System' in J Brunnée, M Doelle and L Rajamani (eds), *Promoting Compliance in an Evolving Climate Regime* (Cambridge University Press, 2012) 102, 111.

116 Brunnée, 'Climate Change and Compliance and Enforcement Processes', above n 2, 306.

117 Doelle, 'Experience with the Facilitative and Enforcement Branches of the Kyoto Compliance System', above n 115, 112.

2.5 Review in the Kyoto compliance system

Unlike its counterparts in the Montreal Protocol and CITES,[118] the Kyoto compliance system incorporates a limited avenue for review. As previously noted, a Party can appeal to the CMP against a decision of the Enforcement Branch if it believes it has been denied due process and the decision 'relates to' Article 3(1) of the Kyoto Protocol.[119] This indicates that Parties do not have a right to appeal against *all* Enforcement Branch decisions, including on the grounds of other legal and technical errors. If the CMP agrees by a three-fourths majority vote that there has been a lack of due process, it can 'override' the decision and the matter will be referred back to the Enforcement Branch.[120] It is significant that the CMP, a political organ, does not have the authority to make a substantive decision on compliance or to overrule such a decision made by the ostensibly independent experts in the Compliance Committee.[121] The avenue of review to the CMP is deliberately limited in order to minimize the possibility of political interference with the decision making of the Enforcement Branch,[122] which is an unusual limitation on the power of states in IEL. Such constraints on political influences are, however, consistent with procedurally constrained decision making.

This avenue for review to the CMP is yet to be tested. Although Croatia initiated an appeal against a final decision of the Enforcement Branch in January 2010, it withdrew the appeal in August 2011 without providing reasons.[123] The compliance procedures and mechanisms under Decision 27/CMP.1 do not specify that the Compliance Committee's duty to make information available to the Party concerned[124] includes a duty to disclose relevant procedural issues, such as potential conflicts of interest, that would constitute grounds for review.[125] Therefore, in practice, it seems likely that this appeals route will continue to serve a largely symbolic function.

In sum, the institutional designers of the Kyoto compliance system attempted to insulate all tiers of its compliance hierarchy from undue political influence – through review teams and Facilitative and Enforcement Branches composed of experts, and by limiting the accountability of these experts to the Parties in the

118 See Chapters 4 and 6.
119 Decision 27/CMP.1, above n 4, section XI, para 1.
120 *Ibid* section XI.
121 UNFCCC Secretariat, 'Procedural Requirements and the Scope and Content of Applicable Law for the Consideration of Appeals under Decision 27/CMP.1', above n 65, para 37. However, the CMP does have authority to change the substantive rule upon which the non-compliance is based: *id*.
122 Lefeber and Oberthur, 'Key Features of the Kyoto Protocol's Compliance System', above n 31, 85.
123 Decision 14/CMP.7, above n 100, para 1.
124 Decision 27/CMP.1, above n 4, section VII, paras 4 and 5; section VIII, para 7; and section IX, para 6.
125 UNFCCC Secretariat, 'Procedural Requirements and the Scope and Content of Applicable Law for the Consideration of Appeals under Decision 27/CMP.1', above n 65, para 31.

CMP. The Compliance Committee has demonstrated potential for procedurally constrained decision making and latent potential for procedural regard. Regrettably, the benefits stemming from these opportunities have been curtailed by their infrequent use in practice.

3 A vignette of compliance decision making: Greece

The first question of implementation brought before the Compliance Committee in 2008 concerned Greece. This non-compliance proceeding constitutes an apt vignette as it highlights two salient issues in the Kyoto compliance system's first commitment period: (1) the perception that ERT processes lacked consistency and transparency and (2) the bypassing of the Facilitative Branch in relation to matters that appear to be squarely within its remit. As I argued in the preceding sections, these two issues – the dual technical and political roles of ERTs, and the circumvention of the Facilitative Branch – have broader implications for the effectiveness of the roles played by administrative procedures in constraining politics and promoting procedural regard in the Kyoto compliance system.

Substantively, the non-compliance issue concerning Greece revolved around problems with knowledge transfer. On 29 December 2006, Greece submitted its initial report required under Article 7 to the Secretariat. The ERT's Initial Review Report for Greece was published approximately one year later on 28 December 2007, and contained a question of implementation in relation to the national system requirements of the Kyoto Protocol, which triggered non-compliance proceedings. The context for these proceedings was that the National Observatory of Athens had played the role of 'technical consultant' in establishing Greece's national system, and this role was in the process of being transferred to the National (Metsovian) Technical University of Athens (NTUA) when the ERT in-country review[126] occurred. The ERT report summarized their factual findings underpinning the question of implementation as follows:

> In particular, the ERT concludes that the maintenance of the institutional and procedural arrangements; the arrangements for the technical competence of the staff; and the capacity for timely performance of Greece's national system is an unresolved problem, and therefore lists it as a question of implementation.[127]

The fact that the timing of the ERT's in-country review coincided with Greece's transition period heightened concerns in the ERT about the capacity of the

126 The experts who conducted the review were representatives from Switzerland, Slovenia, Finland, China, Argentina and the Philippines, achieving the aims of geographical balance among ERT members specified in the rules and procedures: Compliance Committee, Report of the Review of the Initial Report of Greece, CC/ERT/IRR/2007/29, 31 Dec 2007, para 1.

127 *Ibid* para 244.

government officials who would henceforth be responsible for the national system.[128]

This was the first question of implementation raised by an ERT in the Kyoto compliance system, and an arm's length and relatively formal decision-making process appears to have been followed. The ERT report states:

> In accordance with the guidelines for review under Article 8 of the Kyoto Protocol (Decision 22/CMP.1), a draft version of this report was communicated to the Government of Greece, which provided comments that were considered and incorporated in this final version of the report as appropriate.

In this instance, therefore, there is little evidence of the 'considerable give and take' between the ERT and the state under review that characterized subsequent review processes described above.

On 31 December 2007, the ERT submitted their review report for Greece to the Secretariat, triggering the non-compliance process. The Bureau subsequently allocated the matter to the Enforcement Branch,[129] despite the fact that Greece's problems with knowledge transfer appear to have been within the remit of the Facilitative Branch, which is tasked with, inter alia, facilitating, implementing and promoting Annex I Parties' compliance with Protocol commitments that do not relate to emission-reduction commitments.[130] Moreover, the Facilitative Branch is intended to provide an early warning function for potential non-compliance in relation to emissions targets,[131] which again seems appropriate in the light of Greece's transition problems. Nonetheless, on 22 January 2008, the Enforcement Branch agreed to proceed with the matter against Greece.[132] Soon thereafter, Greece was informed of the decision to proceed, and made a written submission on 26 February 2008 requesting a hearing in accordance with the due-process opportunities afforded by the compliance procedures.[133]

In exercising its juridical participation right to make a written submission, Greece criticized the perceived heavy-handedness of the ERT's approach, and the *lack of* facilitative 'give and take' in this instance, which they viewed as inconsistent with ERTs' treatment of similar implementation issues in other Annex I countries. Greece articulated these concerns at the March meeting that informed the

128 Compliance Committee, Written Submission of Greece under Section X of Annex to Decision 27/CMP.1, CC-2007-1-5/Greece/EB, 26 Feb 2008.
129 Compliance Committee, Decision on Preliminary Examination (Party Concerned: Greece), CC-2007-1-2/Greece/EB, 22 Jan 2008, paras 1–2.
130 Ulfstein and Werksman, 'The Kyoto Compliance System', above n 14, 45.
131 Decision 27/CMP.1, above n 4, section IV, para 6.
132 Compliance Committee, Decision on Preliminary Examination, above n 129, para 6.
133 Decision 27/CMP.1, above n 4, annex, section IX, para 2; and Decision 4/CMP.2, above n 45, rule 9(1).

Enforcement Branch's preliminary findings,[134] and reiterated them at length in their written response to these preliminary findings.[135] The nub of these concerns was Greece's claim of 'an absence of an objective methodology' underpinning ERT reviews, leading to inconsistent consequences for similar compliance issues, and placing Greek industry in an 'unfair position' vis-à-vis other Annex I countries.[136]

In its written submission, Greece attempted to substantiate these claims by pointing to numerous examples of problems in other countries' national systems documented in ERT reports that did not lead to questions of implementation for these countries.[137] As Greece noted in its submission, the only requirement for these countries was 'to provide additional updated information on the national system' in the following year's submission.[138] As previously mentioned, these types of consistency concerns have been echoed on numerous occasions by the Compliance Committee as well.[139] In this regard, it is unclear whether the 'respective capacities' of CEITs versus developed economies were informally influencing ERTs' differentiated review responses. This underscores the desirability of these types of issues being dealt with by the procedurally constrained Facilitative Branch, with its extensive due-process guarantees and mandate to take into account the principle of 'common but differentiated responsibilities and respective capacities'.[140]

The Enforcement Branch held a formal hearing on 4–5 March 2008, which provided Greece with a right to be heard. Prior to the hearing, the Enforcement Branch officially requested technical expert advice to complement their own predominantly legal expertise. Four experts were invited from Ghana, Switzerland, Finland and Moldova,[141] mirroring the efforts to achieve balanced participation from developing and developed states reflected in the rules regarding the composition of ERTs and the Compliance Committee. Beyond the non-decisional participation by experts at the hearing, there was no other participation by non-state actors or non-parties.[142] Much of the meeting was open to the public. However, the deliberations on the preliminary finding were held in private,[143] which is consistent with the transparency and participation aims reflected in the rules and procedures discussed above.

134 Compliance Committee, Further Submission of Greece under Section X, Decision 27/CMP.1, CC-2007-1-7/Greece/EB, 9 Apr 2008, 3, para A.
135 *Ibid* 2–3.
136 *Ibid* 3–4.
137 *Ibid.*
138 *Ibid* 3.
139 See Chapter 5, section 2.1.2.
140 Decision 27/CMP.1, above n 4, section IV, para 4.
141 Compliance Committee, Expert Advice: Greece, CC-2007-1-3/Greece/EB, 8 Feb 2008, para 3.
142 Doelle, 'Compliance and Enforcement in the Climate Change Regime', above n 38, 174.
143 UNFCCC Webcast, 'The Third Meeting of the Enforcement Branch: Bonn, 4 March – 6 March 2008' (2008) <http://unfccc.int/resource/webcast/compliance/enf3/templ/ovw_unfccc7729.html>; Doelle, 'Compliance and Enforcement in the Climate Change Regime', above n 38, 175.

The Enforcement Branch adopted a preliminary finding of non-compliance the day after the hearing. The reasons provided by the Enforcement Branch in the preliminary judgment were somewhat limited. The crux of the Enforcement Branch's reasoning is reflected in paragraph 16:

> The information submitted and presented has not been sufficient for the enforcement branch to conclude that the question of implementation has now been fully resolved. Additional information is required that specifically addresses whether and how the national system is maintained through transitions. The enforcement branch agrees with the expert advice provided that a further in-country review of Greece's new national system, in conjunction with a review of an annual inventory report generated by this national system, is required for the enforcement branch to assess present compliance with the guidelines.[144]

These reasons evidence the Enforcement Branch's concern about Greece's ongoing and *future* compliance with the Kyoto Protocol, including through periods of transition, rather than simply evaluating Greece's compliance at the time of the March hearing. Again, this raises questions about whether these issues could have been more appropriately dealt with by the Facilitative Branch given its mandate to provide advice and facilitate implementation, taking into account countries' individual circumstances.[145]

Greece exercised its right to respond to the Enforcement Branch's preliminary findings by making further submissions on 8 April 2008. These submissions argued that, regardless of the transition difficulties at the time of the first in-country ERT review, the responsibility for Greece's national system had now been successfully transferred to the National (Metsovian) Technical University of Athens.[146] Greece also stated that it had submitted a new national emissions inventory to the Secretariat,[147] and argued that it was therefore unnecessary to delay Greece's access to the flexibility mechanisms pending a second in-country ERT review.[148] As noted above, this additional written submission also articulated Greece's concerns about the apparent arbitrariness and inconsistency of the ERT process.[149]

Greece's second written submission was considered in detail by the Enforcement Branch at a second hearing on 16–17 April 2008,[150] but did not alter the

144 Compliance Committee, Preliminary Finding (Party Concerned: Greece), CC-2007-1-6/Greece/EB, 6 Mar 2008, para 16.
145 Decision 27/CMP.1, above n 4, section IV, para 4.
146 Compliance Committee, Further Submission of Greece, above n 134, 2.
147 *Ibid.*
148 *Ibid* 1–4.
149 *Ibid* 2–4.
150 Doelle, 'Compliance and Enforcement in the Climate Change Regime', above n 38, 178–179.

Enforcement Branch's preliminary view. After considering the arguments raised in this second submission, the Enforcement Branch adopted a final decision to confirm its preliminary finding.[151] Unlike the preliminary finding, the final decision was not unanimous.[152] There were no reasons provided for the dissenting vote, which would have illuminated why this member's position had changed since the preliminary finding.[153]

The non-compliance consequences that applied to Greece are consistent with the Enforcement Branch's 'automatic review approach' for different types of non-compliance.[154] As the Enforcement Branch deemed that the compliance issues concerning Greece pertained to methodological and reporting requirements, and eligibility to participate in the flexibility mechanisms, the following consequences were applied:

a. Greece is declared to be in non-compliance.
b. Greece shall develop a [compliance action plan] and submit it within three months to the enforcement branch in accordance with paragraph 2 of section XV [of Decision 27/CMP.1). ...
c. Greece is not eligible to participate in the [flexibility] mechanisms under Articles 6, 12 and 17 of the Protocol pending the resolution of the question of implementation.[155]

Hence, an automatic and non-discretionary response was applied, which is consistent with the rules and procedures for non-compliance,[156] and aligns with accountability aims of predictability and consistency.

A second in-country review involving dialogue between the ERT and Greece appears to have been critical to the ultimate resolution of this matter. The first compliance action plan submitted by Greece on 16 July 2008 was deemed to be inadequate by the Enforcement Branch as it did not address elements and issues stipulated in the final decision.[157] Shortly after the Enforcement Branch asked Greece to submit a revised compliance action plan addressing these issues in October 2008, an ERT conducted a second in-country review in Greece. One of the chief reasons provided in the Enforcement Branch's subsequent decision that

151 Compliance Committee, Final Decision (Party Concerned: Greece), CC-2007-1-8/Greece/EB, 17 Apr 2008, paras 1–3.
152 *Ibid* 2.
153 Compliance Committee, Preliminary Finding, above n 144, para 19.
154 Brunnée, 'Climate Change and Compliance and Enforcement Processes', above n 2, 306.
155 Compliance Committee, Final Decision (Party Concerned: Greece), above n 151, para 5, referring to Compliance Committee, Preliminary Finding (Party Concerned: Greece), CC-2007-1-6/Greece/EB, 6 Mar 2008, para 18.
156 Decision 27/CMP.1, above n 4, section XV, para 5(a), (b) and (c).
157 Compliance Committee, Decision on the Review and Assessment of the Plan Submitted under Paragraph 2 of Section XV (Party Concerned: Greece), CC-2007-1-10/Greece/EB, 7 Oct 2008, paras 3–4.

the question of implementation involving Greece was resolved was the ERT finding from this second in-country review that 'the national system of Greece is performing its required functions, as set out in the annex to decision 19/ CMP.1'.[158] The Enforcement Branch's reasons further highlight the importance of the ERT's dialogue with relevant personnel in Greece, alongside their technical review functions, during the in-country review:

> The review report also confirmed that the ERT had *in-depth discussion on all aspects of the national system with the relevant staff*, and that the transfer of information and data from the institution with previous technical responsibility for the inventory preparation to the new team has been completed.[159]

Thus, ERT review processes were pivotal to instigating and resolving the question of implementation involving Greece.

The non-compliance proceedings concerning Greece illustrate that, in the early stages of the operation of the Kyoto compliance system, there was extensive reliance on both rules and decisions of an administrative character. However, the blurred lines between ERTs' roles as technical experts and compliance facilitators were already evident in Greece's complaints about the lack of consistency and comparability in ERTs' treatment of different countries' ostensibly analogous non-compliance issues. This raises questions about ERTs' suitability to fulfill dual technical and political roles. It also highlights the desirability of greater involvement by the Facilitative Branch which, due to its clear and detailed due-process guarantees, is well suited to managing the procedural and facilitative imperatives of compliance with the Kyoto Protocol.

4 Conclusion

The Kyoto compliance system has developed a complex administrative apparatus with highly developed rules and procedures for accountability, transparency, participation, reason giving and review. On paper, the design of the Facilitative and Enforcement Branches promotes procedurally circumscribed decision making and opportunities for regard for affected states' interests. In practice, however, procedural regard aims have been hampered by the bypassing of the Facilitative Branch and the pivotal, yet weakly constrained, facilitative roles assumed by the expert review teams. These small groups of experts have created a bottleneck limiting the number of cases heard by the Enforcement Branch, which does not appear to reflect the full extent of Annex I Parties' compliance issues.

Attempts to insulate the Compliance Committee's processes from overt political influences have been largely effective, yet the benefits of such constraints are

158 Compliance Committee, Decision under Paragraph 2 of Section X (Party Concerned: Greece), CC-2007-1-13/Greece/EB, 13 Nov 2008, para 7.

159 *Ibid* para 8 (emphasis added).

undermined if decision making occurs elsewhere. The proceduralized forums provided by the Facilitative and Enforcement Branches have not been well utilized due to the movement of regulatory authority to more informal negotiating spaces, which lack due-process guarantees. In particular, due to the relative opaqueness of ERTs' facilitative roles and processes, the extent to which they afford procedural regard to affected states is unclear. These inter-related findings simultaneously underscore both the potential benefits and limitations of proceduralization in global environmental governance.

6 The CITES compliance system and the risks associated with inadequate procedural safeguards

1 Introduction

Unlike later MEAs, such as the Montreal and Kyoto Protocols that have designated compliance bodies and processes, the CITES[1] compliance system is composed of secondary rules that have accumulated in an ad hoc manner since the mid-1980s.[2] The compliance procedures and practices were only compiled and codified in the non-binding *Guide to CITES Compliance Procedures* in 2007.[3] This context provides fertile ground for examining whether the evolution of secondary compliance rules over time has contributed to greater procedural constraints and regard for affected states' interests.

Notably, any such shift within this compliance system has occurred against a backdrop of politically charged North–South tensions shaping CITES regulation of trade in endangered species. The primary, but not exclusive,[4] direction of the commercial exchange of wildlife and derivative products, which is characterized by a 'luxury orientation', is 'South-to-North',[5] meaning that wildlife and wildlife products constitute a source of valued export income for many developing

1 *Convention on International Trade in Endangered Species of Wild Fauna and Flora 1973*, opened for signature 3 March 1973, 993 UNTS 243 (entered into force 1 July 1975). As of May 2017, there were 183 Parties to the Convention, which is one of the highest membership rates of any international conservation agreement: CITES, 'What is CITES?' (2017) <www.cites.org/eng/disc/what.php>.

2 R Reeve, 'The Convention on International Trade in Endangered Species of Wild Fauna and Flora' in G Ulfstein, T Marauhn and A Zimmermann (eds), *Making Treaties Work: Human Rights, Environment and Arms Control* (Cambridge University Press, 2007) 134, 136.

3 CITES, CITES Compliance Procedures, Resolution Conf 14.3, Jun 2007, Annex ('CITES compliance procedures'). See also Chapter 3, section 4.1.3.

4 For example, Montazeri claims that China is 'the final destination of an estimated seventy percent of the world's ivory', with reports suggesting that 'ninety percent of ivory in China is illegal': S Montazeri, 'Protecting the Pachyderm: The Significance of Ivory Trade Regulation for African Elephant Conservation' (2013) 22 *Cardozo Journal of International & Comparative Law* 121, 127.

5 P H Sand, 'Whither CITES? The Evolution of a Treaty Regime in the Borderland of Trade and Environment' (1997) 8 *European Journal of International Law* 29, 29–30.

countries.[6] Paradoxically, despite the strong Northern demand for wildlife products, powerful Northern actors, including both states and influential non-governmental organizations (NGOs), have historically been strident advocates for species protection and preservation, which at times conflict with Southern interests in wildlife export income.[7] These political tensions provide important context for understanding the roles played by administrative procedures in the CITES compliance system. Significantly, this case study demonstrates the risk that, without adequate procedural safeguards, compliance decisions will be influenced by differences in power between states. This risk is exemplified by sanctions for non-compliance that overwhelmingly target developing range states, but not developed export countries. This finding reinforces the importance of robust administrative procedures to ameliorate such risks.

Nonetheless, there are lessons regarding the conditions that support procedural regard that can be distilled from this analysis. This case study demonstrates that a pluralist conception of the appropriate responses to global regulatory problems can be a favorable condition for enhancing procedural regard.[8] Such pluralism came to the fore within the regime in the 1990s, when the paradigm of conservation through sustainable use increasingly gained traction alongside the traditional preservationist philosophy favored by powerful developed states and NGOs. Arguably, the preamble to CITES, which states that 'wild flora and fauna in their many beautiful and varied forms are an irreplaceable part of the natural systems of the earth which must be protected for this and the generations to come', reflects preservationist rationales.[9] However, this agenda, which was influential during the early years of CITES and is underpinned by trade-restrictive policies, has been criticized for disregarding the livelihood strategies of local communities in the range states of wildlife.[10]

6 *Ibid* 30.

7 R Duffy, 'Global Environmental Governance and North–South Dynamics: The Case of CITES' (2013) 31 *Environment and Planning C: Government and Policy* 222.

8 This appears to support Hovell's contention that procedural regard thrives on a pluralistic recognition and appreciation of actors' diverse preferences and interests, which lacks compatibility with regimes predicated on an ostensibly neutral consensus: D Hovell, *The Power of Process: The Value of Due Process in Security Council Sanctions Decision-Making* (Oxford University Press, 2016) 74–76, 138. See also N Krisch, 'Global Administrative Law and the Constitutional Ambition' in P Dobner and M Loughlin (eds), *The Twilight of Constitutionalism* (Oxford University Press, 2010) 245, 254.

9 CITES, above n 1, preamble.

10 J O Velazquez Gomar and L C Stringer, 'Moving Towards Sustainability? An Analysis of CITES' Conservation Policies' (2011) 21 *Environmental Policy and Governance* 240, 240.

In the early 1990s,[11] the artificial consensus on preservation[12] was disrupted by polarizing debates within the regime between advocates of preservation and conservation based on sustainable use. The latter paradigm recognizes that in appropriate institutional settings, 'people's use of wild resources can generate both the motivation and the revenue for conservation and sustainable management of species and their habitats'.[13] This approach to sustainable wildlife utilization is arguably consistent with the Convention text, which seeks to co-ordinate trade in species that are at risk of becoming endangered.[14] Since the 1990s, conservation paradigms have become increasingly influential within the regime, evidenced by Parties' approval of sustainable-use methods.[15] Thus, despite the protectionism evident in the treaty's objects clause, 'the prevailing *zeitgeist* of the early twenty-first century is more that of sustainable use'.[16] The following analysis shows that as competing perspectives regarding conservation were debated and came to be accommodated within the regime, there was greater receptivity to the desirability of procedural regard for affected states' interests in listing decisions, and in non-compliance proceedings.

This chapter traces a gradual shift toward increasing reliance on procedural tools of an administrative character in the CITES compliance system over time,

11 One of the earliest steps taken within CITES to recognize range states' livelihood interests in sustainable use occurred at COP 8 (Kyoto, 1992) with the passing of Resolution Conf 8.3 (Recognition of the Benefits of Trade in Wildlife), which acknowledged that 'commercial trade may be beneficial to the conservation of species and ecosystems and/or to the development of local people when carried out at levels that are not detrimental to the survival of the species in question': CITES, Recognition of the Benefits of Trade in Wildlife, Resolution Conf 8.3 (Rev COP 13), Oct 2004.

12 Couzens makes the point that sustainable utilization was traditionally the dominant paradigm of the 'world's conservation establishment', but went 'underground' for some years: E Couzens, *Whales and Elephants in International Conservation Law and Politics: A Comparative Study* (Routledge, 2014) 134.

13 R Cooney and M Abensperg-Traun, 'Raising Local Community Voices: CITES, Livelihoods and Sustainable Use' (2013) 22(3) *Review of European Community and International Environmental Law* 301, 302.

14 The Convention falls short of explicitly endorsing a balancing of competing interests between environmental protection and development. Cf the principle of sustainable development articulated in the Brundtland Report and Agenda 21: World Commission on Environment and Development, 'Report of the World Commission on Environment and Development: Our Common Future' (1987) <www.un-documents.net/wced-ocf.htm>; UNEP, 'Agenda 21' (1992) <https://sustainabledevelopment.un.org/content/documents/Agenda21.pdf>.

15 See, eg, D M Ong, 'The Convention on International Trade in Endangered Species (CITES, 1973): Implications of Recent Developments in International and EC Environmental Law' (1998) 10(2) *Journal of Environmental Law* 291. Wijnstekers, a former Secretary-General of CITES, observed in 2007 that CITES 'has learned to balance conservation and sustainable use': IISD Reporting Services, 'Summary of the Fourteenth Meeting of the Conference of the Parties to the Convention on International Trade in Endangered Species of Wild Fauna and Flora' (18 June 2007) 21(61) *Earth Negotiations Bulletin* 2 <http://enb.iisd.org/download/pdf/enb2161e.pdf>.

16 Couzens, *Whales and Elephants*, above n 12, 126.

particularly in some domains, revealing a number of key trends. First, as the adoption of administrative procedures enhanced constraints upon politics in some domains, the procedural safeguards in other domains were loosened, undermining the net gains associated with proceduralization. Second, the risk that administrative procedures may serve to perpetuate pre-existing power imbalances appears to be reflected in the apparent bias against weaker states in the CITES compliance system. However, the participation of civil-society representatives and the emergence of a pluralist understanding of the appropriate regulatory responses to achieve the regime's goals have created a more favorable environment for procedural regard. A vignette of compliance decision making involving the United Arab Emirates in section 3 crystallizes this chapter's thematic analysis.

2 Evaluating proceduralization in the CITES compliance system

As previously mentioned, a distinctive feature of the CITES compliance system is that it has evolved through the accretion of secondary rules and practices.[17] Accordingly, the following evaluative process adopts a historical perspective, and focuses upon key developments in the compliance system's shifts toward, or away from, proceduralization over time. Procedures for accountability, transparency and participation have prominent roles in the CITES compliance system, while reason giving and review mechanisms are not foregrounded to the same extent.

2.1 Accountability in the CITES compliance system

2.1.1 Accountability and the Secretariat

The CITES Secretariat plays a pivotal role in the administration of the Convention and its compliance system, and at times pursues its own normative agenda.[18] There are two dimensions of accountability that arise in relation to the Secretariat: the first concerns the extent to which the Secretariat helps to hold states to account for their CITES commitments, and the second concerns the extent to which the Secretariat itself is accountable to the Parties.[19] In relation to the former, the Secretariat's exercise of discretion in escalating non-compliance matters has contributed to a pattern in which less powerful states are the primary targets of compliance proceedings and trade sanctions.[20] For example, while in 1999–2002 Afghanistan

17 Reeve, 'The Convention on International Trade in Endangered Species of Wild Fauna and Flora', above n 2, 136.

18 There is increasing recognition that MEA Secretariats may be 'actors in their own rights': see S Bauer, P-O Busch and B Siebenhüner, 'Treaty Secretariats in Global Environmental Governance' in F Bierman, B Siebenhüner and A Schreyögg (eds), ***International Organizations in Global Environmental Governance*** (Routledge, 2009) 174, 174–175.

19 See Chapter 2, section 3.1.

20 P H Sand, 'Enforcing CITES: The Rise and Fall of Trade Sanctions' (2013) 22(3) *Review of European Community & International Environmental Law* 251, 261; P H Sand, 'International

and Rwanda were subject to trade suspensions for persistent failure to designate appropriate Scientific Authorities, the Secretariat has never recommended similar action against Japan for its failure to designate separate Management and Scientific Authorities under Article IX.[21] This issue, which reflects and reinforces a lack of regard for less powerful states' interests, is explored in the following sub-section addressing accountability and the Standing Committee. This sub-section focuses on the second dimension of accountability noted above. In this regard, the use of oversight mechanisms has contributed to disciplining unilateral tendencies within the Secretariat, aligning with desirable aims associated with procedurally constrained decision making.

The CITES Secretariat has on numerous occasions acted unilaterally and without the support of the Parties,[22] raising concerns about its accountability. For example, at COP 7 in Lausanne, Switzerland in October 1989, preservationist NGOs that were lobbying strongly for banning the ivory trade through the listing of the African elephant on Appendix I accused the Secretariat of partiality on this issue. This issue has compliance implications as the Asian elephant had already been listed on Appendix I in 1975, and if the African elephant was also listed on Appendix I, a ban on all international trade in ivory would be effectuated, extending the grounds for non-compliance. Specifically, NGOs raised concerns about the Secretariat's role in, inter alia, lobbying in the media *against* the Appendix I listing.[23] It appears that the Secretariat was acting without the Parties' support in this regard, and after fraught debates, the COP did in fact approve the listing of African elephants in Appendix I.[24] As may be expected in instances when the positions of the Secretariat and the Parties diverge,[25] the Parties' preferences prevailed through the COP vote. It is notable that the COP decision to list the African elephant on Appendix I in the late 1980s reflected a preservationist agenda, and did not accommodate affected African states' interests in sustainable use.[26]

In response to these events, additional procedural constraints were placed on the Secretariat. In terms of the Secretariat's accountability to the Parties, the Parties may be seen to have imposed 'consequences' – an element of accountability – upon the Secretariat at COP 7. This was achieved by adding two paragraphs to prior

Protection of Endangered Species' (2017) 20 *Asia Pacific Journal of Environmental Law* (forthcoming); R Reeve, *Policing International Trade in Endangered Species: The CITES Treaty and Compliance* (Earthscan, 2002) 324.

21 Sand argues that the dual scientific and management functions of the Japan Fisheries Agency 'squarely contravenes the mandatory criteria for designation of scientific authorities': Sand, 'Enforcing CITES', above n 20, 261–262.

22 Reeve, *Policing International Trade in Endangered Species*, above n 20, 258–266.

23 *Ibid* 259.

24 CITES, Trade in Elephant Specimens, Resolution Conf 10.10 (Rev COP 17), Oct 2016.

25 A tension may arise here as Secretariats of MEAs are required to serve the interests of the treaty and the Parties, which may at times conflict: P Kishore, 'A Comparative Analysis of Secretariats Created under Select Treaty Regimes' (2011) 45 *The International Lawyer* 1051, 1053.

26 Couzens, *Whales and Elephants*, above n 12, 131–139.

resolutions that strengthened the Standing Committee's oversight powers of the Secretariat's budget processes, including the requirement of review before the Secretariat accepts funds from non-governmental sources, such as wildlife trade groups.[27] This suggests that the Parties shared the NGOs' concerns about irregularities in the Secretariat's operations and the risk of special interest groups' undue influence.[28] The Standing Committee took additional steps to hold the Secretariat to account by authorizing an independent inquiry in 1997–1998.[29] Following this inquiry and a further investigation conducted by the United Nations Office of Internal Oversight, in June 1998 the Executive Director of UNEP informed the Standing Committee that the contracts of two suspended officers were to end, and that the Secretary-General would be assigned to other duties.[30]

Thereafter, the Secretary-General's replacement announced to the Standing Committee his intention to address the Secretariat's accountability issues:

> One of the first things to do therefore is to complete the re-establishment of the Secretariat as a strong, confident, professional and efficient entity in which you are able to believe. This ... requires an attitude and commitment from the staff, commensurate with their position as United Nations officials. ... We need to urgently restore the confidence of the Parties – and of the Standing Committee in particular – in the Secretariat.[31]

The Secretariat has not continued to be plagued by the dramatic types of accountability issues it experienced during the 1990s.[32] This shift suggests that the use of accountability and oversight mechanisms helped to curb the Secretariat's unilateral tendencies, and that there has been an improvement in the Secretariat's adherence to decision-making protocols since this time.

2.1.2 Accountability and the Standing Committee

In ensuring states' accountability for their CITES obligations, does the Standing Committee apply consequences consistently with pre-determined standards, taking into account the interests of affected states? As this sub-section shows, it is

27 CITES, Financing and Budgeting of the Secretariat and of Meetings of the Conference of the Parties, Resolution Conf 7.2, Oct 1989 (Rev COP 12), Nov 2002.

28 According to Meredith, two-thirds of the Secretariat's budget from 1985 to 1989 came from 'donations' from ivory dealers: M Meredith, *Africa's Elephant: A Biography* (Hodder and Stoughton, 2001) 211–214.

29 As discussed in Chapter 2, section 3.1, independent inspections and audits of the internal operations of MEA compliance systems can further enhance accountability.

30 CITES, Matters Relating to the Standing Committee: Report of the Chairman, prepared by the Secretariat for COP 11, Doc 11.8, Apr 2000, para 28.

31 CITES, 'CITES World: Official Newsletter of the Parties' (June 1999) Issue 3, 2 <www.cites.org/eng/news/world/3.pdf>.

32 This conclusion is informed by perusal of the Standing Committee reports available on the CITES website: CITES, 'The Standing Committee' (2017) <www.cites.org/eng/com/sc/index.php>.

concerning that the CITES Standing Committee's record on compliance indicates favorable treatment for developed states. The accountability procedures adopted have not been successful in disciplining power politics or promoting procedural regard in this instance.

The CITES compliance procedures, which were only adopted in 2007, reflect elements of the principle of common but differentiated responsibilities (CBDR) that rose to prominence in IEL during the 1990s and early 2000s.[33] In particular, paragraph 32 of the compliance procedures notes that, when the Standing Committee is deciding upon which measure is appropriate, it should take into account a range of factors, including: 'a) The capacity of the Party concerned, especially developing countries, and in particular the least developed and small island developing States and Parties with economies in transition …'

Sub-paragraph (a) reflects an approach to compliance that pays regard to the interests of less powerful states. Despite this recommended approach, there is little evidence of the capacity of affected countries being taken into consideration in compliance decisions in practice. This lack of regard is compounded by the frequent use of trade-suspension measures without prior recourse to the other, less intrusive measures recommended in the CITES compliance procedures.[34] Of course, trade sanctions are not designed to take into account local livelihoods in range state countries derived from sustainable species utilization.[35]

The non-compliance consequences imposed in the last three decades do not align with non-discriminatory accountability aims or favorable differential treatment for developing states. Rather, these consequences appear to be skewed toward powerful countries' interests. Of the 60 Parties and non-Parties that have been subject to trade embargoes since 1985, 58 (97%) have been developing countries.[36] As Sand observes, this comes as something of an 'empirical surprise' as it may be anticipated that compliance gaps and loopholes would be evident in export, transit and import countries, rather than just in export-oriented developing countries.[37] The Secretariat has acknowledged that the majority of Parties subjected to trade recommendation suspensions are range states or source developing countries.[38] Indeed, it has recently recommended that the COP should instruct the Standing Committee to explicitly ensure that compliance approaches include 'source, transit

33 Rajamani charts the rise of the principle of CBDR in the decade between the first Rio Earth Summit in 1992 and the World Summit on Sustainable Development in 2002: L Rajamani, 'The Changing Fortunes of Differential Treatment in the Evolution of International Environmental Law' (2012) 88(3) *International Affairs* 605, 606. See further Chapter 2, section 5.1.

34 CITES compliance procedures, above n 3, para 30. See further Chapter 3, section 4.1.4.

35 Velazquez Gomar and Stringer, 'Moving Towards Sustainability', above n 10, 251.

36 Sand, 'International Protection of Endangered Species', above n 20; CITES, 'Countries Currently Subject to a Recommendation to Suspend Trade' (last updated 1 Feb 2017) <http://cites.org/eng/resources/ref/suspend.php>.

37 Sand, 'Enforcing CITES', above n 20, 261.

38 CITES, CITES Compliance Matters, prepared by the Secretariat for COP 17, Doc 23, Jul 2016, para 37.

and destination countries in a more integrated manner'.[39] This disproportionate focus on 'easy developing country targets'[40] indicates a discriminatory approach to holding states to account for their international commitments that cuts against the grain of both accountability aims and the CBDR principle.

An example of this trend was evident in the 1980s when the Secretariat's infractions reports[41] repeatedly noted Germany's violations of its CITES commitments, yet sanctions were never recommended. More recently, the Standing Committee appeared to adopt a discriminatory approach to enforcing the mandatory requirements for submission of annual reports by national management authorities.[42] In 2000, Resolution 11.4,[43] which, inter alia, encourages Parties' adherence to the International Convention for the Regulation of Whaling,[44] was adopted. The Japan Fisheries Agency[45] apparently viewed this resolution as '"anti-whaling" and hence hostile'.[46] The Agency abruptly ceased reporting on permits issued for CITES-listed marine species in 2000, and did not resume reporting until 2009.[47] Between 2002 and 2013, 14 other CITES Parties were subject to trade suspensions for failure to submit annual reports for three consecutive years, yet the Japan Fisheries Agency's persistent inadequate reporting over nine years did not attract such an embargo.[48] Behind the United States, Japan is the largest financial contributor to the CITES Trust Fund,[49] and a very long-term[50] and

39 *Ibid.*
40 Reeve, *Policing International Trade in Endangered Species*, above n 20, 324.
41 Sand, 'Whither CITES?', above n 5, 49 fn 120.
42 CITES, National Reports, Resolution Conf 11.17 (Rev COP 17), Oct 2016.
43 CITES, Conservation of Cetaceans, Trade in Cetacean Specimens and the Relationship with the International Whaling Commission, Resolution Conf 11.4 (Rev COP 12), Nov 2002.
44 *International Convention for the Regulation of Whaling 1946,* opened for signature 2 December 1946, 161 UNTS 72 (entered into force 10 November 1948).
45 The Japan Fisheries Agency is a nominated Management Authority for marine species: CITES, 'Japan' (2017) <www.cites.org/eng/cms/index.php/component/cp/country/JP>.
46 Sand, 'Enforcing CITES', above n 20, 262.
47 *Ibid.* It should be noted, however, that the Japanese Ministry of Economy, Trade and Industry, which is the designated Management Authority for *non-marine* CITES-listed specimens, continued to submit reports in accordance with CITES, art VIII(7) during this period: *ibid* fn 121.
48 *Ibid* 262.
49 As of June 2016, Japan had contributed $647,393 for 2016, with the next highest contributor being Germany, with contributions of $426,754: CITES, CITES Trust Fund (CTL) Status of Contributions as of 30 June 2016 (in US Dollars), prepared by the Secretariat for COP 17, Doc 7.3, 8 Aug 2016, Annex 10, Table 1.
50 Japan hosted COP 8 in Kyoto in 1992, and was thus elected as a Standing Committee member as the next and previous host party from 1989 to 1994. Japan was also elected three times as a representative of Asia from 1994 to 2000, 2004 to 2010 and 2010 to 2016, and was the Chair of the Standing Committee from 1994 to 1997: Ministry of Foreign Affairs of Japan, 'The Convention on International Trade in Endangered Species of Fauna and Flora (CITES)' (2016) <www.mofa.go.jp/policy/environment/convention/cites.html>. At COP 17 (Johannesburg, 2016), Japan was succeeded by China as the Standing Committee's full member for the Asian region, with Japan assuming the role of alternate member.

influential member of the Standing Committee. Japan's participation in CITES 'is considered essential because it is one of the world's largest consumers of many wildlife products',[51] raising questions about whether it is receiving favorable differential treatment for diplomatic reasons.[52] To the extent that the Standing Committee is, or is perceived to be, more lenient in its recommendation of trade sanctions toward powerful developed countries, it creates an impression that compliance is strongly shaped by differences in power between the Parties. This signifies both procedural and substantive disregard for developing states' interests.

2.1.3 Accountability and the COP

Thus far, I have argued that administrative procedures have played constructive roles in enhancing the Secretariat's accountability to the Parties, but have had little apparent impact on constraining power politics or enhancing regard for affected states' interests in relation to the compliance consequences recommended by the Secretariat and Standing Committee. This sub-section explores the COP's role in varying and extending the accountability expectations under CITES by developing new, and adjusting existing, compliance obligations through Appendices I and II listing decisions. By shifting and developing the parameters for non-compliance, the COP has the power to change the compliance obligations for which states are accountable. The COP's practices in this regard reflect traditional diplomatic and political decision-making modes, which are not significantly procedurally constrained.

Recommendations of the COP, which since 1994 have been known as 'resolutions', 'revised resolutions' and 'decisions', have been highly influential in the development of the CITES regime and its compliance system. Resolutions are 'intended to provide long-standing guidance', whereas decisions 'contain instructions to a specific committee or to the Secretariat' and therefore have an expiry date.[53] Both these types of instruments have played a pivotal role in the development of the CITES compliance system in the absence of a provision in the Convention specifically authorizing the development of such a system.[54] The COP's power to make such recommendations is authorized under Article XI(3)(e) of the Convention,[55] and the decision-making processes that must be followed

51 P Mofson, 'Protecting Wildlife from Trade: Japan's Involvement in the Convention on International Trade in Endangered Species' (1994) 3 *The Journal of Environment & Development* 91, 95.
52 Sand, 'International Protection of Endangered Species', above n 20.
53 CITES, 'Decisions of the Conference of the Parties to CITES in Effect after the 17th Meeting' (2017) <https://cites.org/eng/dec/index.php>.
54 P G Davies, 'Non-Compliance – A Pivotal or Secondary Function of CoP Governance?' (2013) 15 *International Community Law Review* 77, 95. See further Chapter 3, section 4.1.3.
55 CITES, art XI, para 3(e) states that the COP may 'where appropriate, make recommendations for improving the effectiveness of the present Convention'.

are specified in the COP Rules of Procedure.[56] COP decisions are taken by a two-thirds majority vote and bind all Parties, even those who did not support the decision. From a traditional international law standpoint, there is debate about the legitimacy of delegated law-making by the COP due to its departure from the consent-based structure of treaty law.[57] Nonetheless, four decades of practice within the CITES COP indicates Parties' willingness to honor such decisions, signaling tacit acceptance of their legal status.[58]

The COP is also entrusted to adopt amendments to Appendices I and II, which shape the compliance rules that Parties must follow in accordance with Article XV.[59] The decision to list a species on Appendices I or II either restricts or prohibits trade in that species, thereby creating potential grounds for non-compliance if trade persists. The primacy of state sovereignty is preserved as states may 'opt out' by entering reservations against any COP amendments made to Appendices I and II.[60] The COP has defined biological and trade criteria against which such listing decisions must be made. In more recent times, there has been scope for other concerns, such as the impacts of a listing decision upon the livelihoods of rural communities, to influence decision making. Specifically, at COP 13 (Thailand, 2004), an additional paragraph was added to Resolution Conf 8.3 stating that the COP 'recognises that implementation of CITES-listing decisions should take into account potential impacts on the livelihoods of the poor'.[61] This opportunity for procedural regard is significant as the COP's decision-making processes play a critical role in defining the parameters for non-compliance within the CITES regime, leading to frequently adjusted expectations regarding the conduct for which states must account.

2.2 Transparency in the CITES compliance system

Transparency is an aspiration of CITES, as reflected in the regime's current strategic vision, which lists 'promoting transparency and wider involvement of civil society in the development of conservation policies and practices' as one

56 CITES, Rules of Procedure of the Conference of the Parties, Rev COP 14, Jun 2007 ('COP Rules of Procedure').

57 See, eg, J Brunnée, 'COPing with Consent: Law-Making Under Multilateral Environment Agreements' (2002) 15(1) *Leiden Journal of International Law* 1; R R Churchill and G Ulfstein, 'Autonomous Institutional Arrangements in Multilateral Environmental Agreements: A Little-Noticed Phenomenon in International Law' (2000) 94 *American Journal of International Law* 623, 623.

58 T Gehring, 'Treaty-Making and Treaty Evolution' in D Bodansky, J Brunnée and E Hey (eds), *The Oxford Handbook of International Environmental Law* (Oxford University Press, 2007) 467, 492–495; see also J Brunnée, 'International Legal Accountability Through the Lens of the Law of State Responsibility' (2005) 36 *Netherlands Yearbook of International Law* 21, 53.

59 CITES, above n 1, art XI, para 3(b).

60 *Ibid* art XV, para 3.

61 CITES, Recognition of the Benefits of Trade in Wildlife, above n 11.

of the Convention's priorities.[62] However, the CITES compliance system has a mixed record in relation to transparency. The analysis that follows first outlines gains and losses in the transparent rules and decision making in the Secretariat's compliance-related roles, and argues that the net gains stemming from these changes are questionable. Thereafter, transparency concerns pertaining to the Standing Committee and COP's ongoing preference for secret ballots are canvassed.

2.2.1 Transparency and the Secretariat

The types of accountability challenges faced by the Secretariat in the late 1980s and 1990s detailed above were interlinked with a lack of transparency regarding the Secretariat's operations. Upon assuming the role of Secretary-General in 1999, Willem Wijnstekers nominated the improvement of the Secretariat's transparency record as one of his primary tasks:

> Transparency will be another key word in my activities. I intend to enhance policy transparency through a permanent, well-presented and accessible flow of information from the Secretariat to the Parties and others on activities and results. For that purpose, the Secretariat will need to improve its overall communication capacity and capabilities and make better use of the latest available technologies.[63]

This new commitment to transparency regarding the Secretariat's operations was facilitated by the growth of Internet technologies in the late 1990s and early 2000s.[64] These technologies enabled external transparency gains, such as making preparatory meeting documents, including reports from the Secretariat, and Committee meeting reports for the Standing Committee[65] and COP,[66] available on the CITES website. However, in line with common practice in other MEAs, the Secretariat's internal meetings and deliberations, including in relation to compliance,[67] are not externally transparent.

In addition to these improvements in transparency, in some instances, previous transparency gains have been subsequently eroded. A notable example of this is the

62 CITES, CITES Strategic Vision: 2008–2020, Resolution Conf 16.3 (Rev COP 17), Oct 2016, Annex.

63 W Wijnstekers, *The Evolution of CITES*, 9th edition (2011) 2 <https://portals.iucn.org/library/efiles/documents/CITES-012-2011.pdf>.

64 See generally, D C Esty, 'Environmental Protection in the Information Age' (2004) 79 *New York University Law Review* 115, 167–170.

65 CITES, 'The Standing Committee', above n 32.

66 CITES, 'Conference of the Parties' (2017) <https://cites.org/eng/cop/index.php>.

67 This conforms with a requirement of the CITES compliance procedures, which states that while '[g]enerally, findings, reports and communications in compliance matters are not treated confidentially', 'communications between the Secretariat and individual Parties on specific compliance matters are generally confidential': CITES compliance procedures, above n 3, para 6.

jettisoning of the Secretariat's infractions reports for COP meetings, which from 1985 to 2000 provided 'the only regularly updated publicly available record of national implementation and enforcement of CITES'.[68] At its fifth meeting in Buenos Aires in 1985, the COP agreed that the Secretariat should compile and submit a separate report on 'alleged infractions' for the next and subsequent COP meetings.[69] These infractions reports provided numerous detailed case studies on attempted violations or evasion of Convention provisions,[70] and represented the high water mark of the Secretariat's internal and external transparency on compliance.

Since the late 1990s, transparency in compliance reporting has eroded significantly. At COP 10 in Harare, Zimbabwe in 1997, Decision 10.122 stipulated that the Secretariat should distinguish between alleged infractions relating to the provisions of the Convention and those relating to COP resolutions in its infractions reporting.[71] The next infractions report, presented at COP 11 in Gigiri, Kenya in 2000, demonstrated that the Secretariat had gone significantly beyond the mandate provided in Decision 10.122, and decided only to report: (1) infractions that, in the Secretariat's opinion, were a flagrant or deliberate contravention of either the Convention or resolutions or (2) non-compliance issues that the Secretariat was unable to resolve 'on a bi-lateral basis'.[72] This marked a turning point for infractions reporting, resulting in a marked diminution of publicly available compliance information. For example, while the 'Report on Alleged Infractions' for COP 8 (Kyoto, 1992) ran to 86 pages,[73] the report for COP 9 (Fort Lauderdale, 1994) was 55 pages,[74] the report for COP 10 (Harare, 1997) was 47 pages,[75] the report for COP 11 (Gigiri, 2000) was nine pages,[76] and the report for COP 12 (Chile, 2002) was re-branded as an 'Enforcement Report' and ran to 13 pages.[77] Thus, as various other measures for increasing accountability and transparency were introduced in the late 1990s, the Secretariat decided to reduce transparency in infractions reporting.

68 Reeve, *Policing International Trade in Endangered Species*, above n 20, 73.
69 CITES, Interpretation and Implementation of the Convention – Review of Alleged Infractions, prepared for the Sixth Meeting of the Conference of the Parties Ottawa (Canada), Doc 6.19 (Rev), Jul 1987.
70 See, eg, CITES, Review of Alleged Infractions and Other Problems of Enforcement of the Convention, prepared by the Secretariat for COP 8, Doc 8.19 (Rev), Mar 1992.
71 See Decision 10.122 reported in CITES, Review of Alleged Infractions and Other Problems of Implementation of the Convention, prepared by the Secretariat for COP 11, Doc 11.20.1, Apr 2000.
72 *Ibid* paras 46–47.
73 CITES, Review of Alleged Infractions for COP 8, above n 70.
74 CITES, Review of Alleged Infractions and Other Problems of Implementation of the Convention, prepared by the Secretariat for COP 9, Doc 9.22 (Rev), Nov 1994.
75 CITES, Review of Alleged Infractions and Other Problems of Implementation of the Convention, prepared by the Secretariat for COP 10, Doc 10.28 (Rev), Jun 1997.
76 CITES, Review of Alleged Infractions for COP 11, above n 71.
77 CITES, Enforcement Matters, prepared by the Secretariat for COP 12, COP12 Doc 27, Nov 2002.

The reasons that the Secretariat provided for the decision at COP 11 to focus on high-level discussion of selected egregious infractions included, inter alia, Parties' dissatisfaction with infractions reporting that cast their state in a negative light, and resource constraints making the reporting of multiple incidents 'unmanageable'.[78] While the latter may be a valid practical concern, the former undermines an important deterrent function of transparency. This exemplifies a tension between the Secretariat's role in serving treaty interests by providing comprehensive compliance reporting, and serving the Parties' self-interest in avoiding reputational damage.[79] Here, the Secretariat's efforts to appease the Parties resulted in a dilution of previous transparency gains.

As a result of these changes, there is no longer one comprehensive, publicly available report that is a central repository for non-compliance information. Since COP 12 in Santiago, Chile in 2002, the Secretariat has included in its reports on enforcement matters any relevant information regarding illegal trade in Appendix I or II species, or ineffective implementation of the Convention. Information about other potential compliance matters concerning annual reports, national legislation or the Review of Significant Trade is included in other reports.[80] The introduction of CITES Alerts, which are information bulletins on compliance issues sent to national Management Authorities since 2000,[81] and the computerized Trade Infraction and Global Enforcement Recording System in 2002,[82] partially compensated for internal transparency losses stemming from the above-mentioned changes to infractions reporting. Despite this, the fractured nature of compliance reporting remains an issue. On balance, therefore, the overall gains in transparency in the CITES compliance system since the late 1990s are questionable.

2.2.2 Transparency and the Standing Committee and COP

Like the Secretariat, the Standing Committee and COP have a mixed record on transparency. According to the CITES compliance procedures, 'decisions on whether to close or keep open debates in compliance matters are taken according to the Rules of Procedure of the body considering the matter and generally reasons are given'.[83] While historically most meetings concerning non-compliance by specific Parties were held in closed session, NGO observers were admitted to

78 CITES, Review of Alleged Infractions for COP 11, above n 71, paras 13–14.

79 See Kishore, 'A Comparative Analysis of Secretariats Created under Select Treaty Regimes', above n 25, 1053.

80 CITES, Enforcement Matters (COP 12), above n 77, paras 12–14.

81 However, the Secretariat's ongoing resource constraints hamper its ability to enter all the information it receives into this database: CITES, Interpretation and Implementation of the Convention: Compliance and Enforcement, prepared by the Secretariat for COP 16, Doc 29 (Rev 1), Mar 2013, para 66.

82 CITES, Enforcement Matters, prepared by the Secretariat for COP 16, above n 38, para 2.

83 CITES compliance procedures, above n 3, para 7.

these Standing Committee meetings from October 2006.[84] Since then, the use of closed sessions has become 'quite rare',[85] aligning with transparency goals.

However, the ongoing use of secret ballots in both Standing Committee and COP meetings has the capacity to circumvent the transparency benefits of open meetings. The frequent use of secret ballots for controversial amendments to Appendices and COP resolutions, both of which may have a direct bearing on compliance issues, has been subject to ongoing and politically tense debates in the Standing Committee and COP. The Secretariat has chronicled three different approaches to the use of secret ballots in the COP's procedures since 1976: (1) from COP 1 (Berne, 1976) to COP 5 (Buenos Aires, 1985), if a Party requested a secret ballot, that is how the vote would proceed, (2) from COP 6 (Ottawa, 1987) to COP 8 (Kyoto, 1992), a vote could only be taken by a secret ballot if requested by one Party, seconded by another and then supported by a simple majority and (3) from COP 9 (Fort Lauderdale, 1994) to COP 15 (Doha, 2010), a vote could be taken by secret ballot if requested by one Party, and seconded by ten Parties.[86] In other words, the progress toward greater voting transparency made in the late 1980s and early 1990s has been partially eroded since COP 9 in 1994.

Although Rule 25 of the COP Rules of Procedure states that any representative may request a secret ballot, 'although it shall not normally be used',[87] in practice, secret ballots have not been reserved for exceptional cases. As documented by the Secretariat, 69 votes were taken by secret ballot between COP 10 (Harare, 1997) and COP 15 (Doha, 2010).[88] In the light of this frequent reliance on secret ballots, at COP 16 in Bangkok in 2013, the Parties voted on the issue of reforming Rule 25 of the Rules of Procedure. In the sessions, the issue was debated in a way that highlighted the unresolved question of whether secret ballots insulated countries from susceptibility to political pressures from other states or lobbyists, or whether they served to conceal such political pressures.[89] After a protracted and fractious debate, two formal proposals were voted upon,[90] again using secret

84 Reeve, 'The Convention on International Trade in Endangered Species of Wild Fauna and Flora', above n 2, 142.

85 CITES, 'Participation of Non-State Actors in CITES' (Presentation UNECE/Forest Europe Seminar on Developing an Effective International Legal Instrument, Geneva, Switzerland, 7 Sept 2010) <www.unece.org/fileadmin/DAM/timber/meetings/20100907/6_Marceil_Yeater.pdf>.

86 CITES, Improving Transparency of Voting during Meetings of the Conference of the Parties, prepared by the Secretariat for the 62nd meeting of the Standing Committee, SC62 Doc 10.3, Jul 2012, para 7.

87 COP Rules of Procedure, above n 56, rule 25, para 2.

88 CITES, Improving Transparency of Voting during Meetings of the Conference of the Parties, above n 86, para 6.

89 IISD Reporting Services, 'Summary of the Sixteenth Meeting of the Conference of the Parties to the Convention on International Trade in Endangered Species of Wild Fauna and Flora' (18 March 2013) 21 (83) *Earth Negotiations Bulletin* <www.iisd.ca/vol21/enb2183e.html>.

90 See CITES, Proposed Amendment to Rule 25 on Methods of Voting: Use of Secret Ballots, submitted by Chile and Mexico for COP 16, Doc 4.3 (Rev 1), Mar 2013, and CITES,

ballots. Due to a failure to reach a two-thirds majority agreement on any of the proposed reforms, Rule 25 of the Rules of Procedure on the use of secret ballots remains unchanged and in force.[91] The continued, routinized use of secret ballots is out of step with CITES' own transparency goals[92] and the broader transparency agenda in global environmental governance.[93] It also evidences a preference for diplomatic, political decision making over more transparent processes of an administrative character in this domain.

2.3 Participation in the CITES compliance system

CITES was a forerunner in terms of providing non-decisional participation rights for NGOs in an international environmental treaty.[94] The extensive civil-society participation rights in the CITES COP are a primary focus of the following analysis. It is acknowledged that there is extensive non-decisional participation by NGOs in the Secretariat[95] and Standing Committee[96] as well. However, a focus on COP processes exemplifies two pertinent issues for this analysis: first, the risk that NGOs may reinforce Northern interests in wildlife preservation at developing range states' expense and, second, the potential of civil-society representatives to act as advocates for developing countries' interests. The discussion in this sub-section concludes with an overview of juridical and decisional participation rights in the Secretariat, Standing Committee and COP.

2.3.1 Non-decisional participation rights and the COP

In its early years, the CITES Convention was unusual in its privileging of NGO participation at COP meetings. This is an approach that has since been replicated

Proposal to Improve Transparency of Voting During Meetings of the Conference of the Parties, submitted by Denmark on behalf of the EU Member States for COP 16, Doc 4.2 (Rev 1), Mar 2013.

91 IISD Reporting Services, 'Summary of the Sixteenth Meeting of the Conference of the Parties', above n 89.

92 See CITES, CITES Strategic Vision, above n 62, 2.

93 See Chapter 2, section 3.2.

94 S Charnovitz, 'Nongovernmental Organizations and International Law' (2006) 100 *American Journal of International Law* 348, 369–370.

95 Article XII(1) of the Convention provides that the Secretariat may be assisted by 'suitable' national, inter-governmental or non-governmental bodies that are 'technically qualified in protection, conservation and management of wild flora and fauna'.

96 Prior to 2003, meetings of the Standing Committee were closed to all except contracted NGOs – the International Union for the Conservation of Nature (IUCN), the United Nations Environment Programme's World Conservation Monitoring Centre (UNEP-WCMC) and TRAFFIC. Since this time, other NGOs with expertise in wildlife issues have been permitted to apply to attend and participate at Standing Committee meetings: R Reeve, 'Wildlife Trade, Sanctions and Compliance: Lessons from the CITES Regime' (2006) 82(5) *International Affairs* 881, 883.

in the UNFCCC and other MEAs.[97] CITES' position on NGO participation, which is perhaps partly explained by the International Union for Conservation of Nature's (IUCN's) role in facilitating the drafting of the Convention,[98] is codified in Article XI(7). In order to be eligible to be registered as an observer under this provision, an organization must demonstrate that it is qualified in the protection, conservation or management of wild fauna and flora, and that it is an organization in its own right with a legal persona and an international character, remit and program of activities.[99] Before speaking at meetings, approved observers must be invited to do so by the Presiding Officer, who is required to give precedence to the delegates and the Secretariat first, and then among observers, to non-Party states, intergovernmental organizations and NGOs in that order.[100] Members of the public are afforded visiting rights to COP meetings, but not participatory rights.[101]

In practice, NGOs are active at COP meetings through verbal interventions, proposing amendments to COP recommendations and contributing to discussions in working groups.[102] There are numerous high-profile environmental NGOs that are particularly influential in shaping COP decision making. One of these is the IUCN, which serves as a forum for 1300 member organizations, more than 200 of which are governments.[103] A second influential NGO is the Trade Records Analysis of Fauna and Flora in Commerce (TRAFFIC), a partnership between the World Wild Fund for Nature (WWF) and the World Conservation Union.[104] TRAFFIC was established in 1976 to, inter alia, assist the CITES compliance bodies in implementing the provisions of the Convention.[105] IUCN and TRAFFIC publish in-depth and influential analyses of, and recommendations for, all the

97 K Raustiala, 'The "Participatory Revolution" in International Environmental Law' (1997) 21 *Harvard Environmental Law Review* 537, 547.

98 *Ibid* 559.

99 CITES, Participation of Observers at Meetings of the Conference of the Parties, Resolution Conf 13.8 (Rev COP 16), Mar 2013.

100 COP Rules of Procedure, above n 56, rule 17, para 2.

101 CITES, 'Conference of the Parties', above n 66.

102 C Fuchs, 'Convention on International Trade in Endangered Species of Wild Fauna and Flora (CITES) – Conservation Efforts Undermine the Legality Principle' (2008) 9(11) *German Law Journal* 1565, 1581.

103 IUCN, 'About' (2017) <www.iucn.org/about>.

104 The World Conservation Union is now known as the International Union for the Conservation of Nature.

105 Utilizing its network of thirty offices worldwide that are organized in seven regional programs, TRAFFIC tracks illicit trade routes, examines underlying causes and demands of such trade and investigates potential smuggling activities: TRAFFIC, 'About Traffic' (2017) <www.traffic.org/overview/>. Abbott et al describe TRAFFIC as an 'intermediary' that helps the CITES Secretariat to monitor state compliance with their CITES commitments: K W Abbott, P Genschel, D Snidal and B Zangl, 'Orchestration: Global Governance through Intermediaries' in K W Abbott, P Genschel, D Snidal and B Zangl (eds), *International Organizations as Orchestrators* (Cambridge University Press, 2015) 3, 5, 23.

formal Party proposals to be discussed at each COP.[106] Thus, CITES inclusive participatory rules are reflected in, and at times even exceeded by, the extent of NGO participation in practice.

Some commentators question the representativeness, accountability and legitimacy of NGO involvement in international decision-making processes.[107] These types of concerns have some purchase in relation to CITES. In particular, there is a perception that the most influential environmental NGOs have, at key historical junctures, represented Northern wildlife preservation interests and sidelined Southern interests in wildlife export income, thereby perpetuating and exacerbating existing global power imbalances.[108] For example, Duffy critiques the participation of influential NGOs in the lead up to the 1989 ivory trade ban discussed above, noting that 'they were not accountable to African communities and governments that lived with and managed elephants'.[109] Such critiques suggest disregard for affected states' compliance interests, although it should be recalled that ultimately it is the COP that makes the final listing decisions.

In more recent times, NGOs have adopted advocacy stances that are more supportive of developing range states' interests in sustainably managing species and generating revenue for conservation. This trend corresponds with a broader shift within the regime to allow greater accommodation of sustainable-use perspectives. For example, at COP 16 in 2013, NGOs were influential in advocating for the continuation of legal means of ivory trade. That is, their roles at COP 16 involved advocating *against* limiting one-off ivory sales and other means of legal ivory trade currently permitted under the Convention.[110] In this instance, NGOs' involvement in CITES served to promote an outcome that was consistent with conservation based on sustainable use, reflecting a stance that is supportive of range states' interests in wildlife export income.

In addition to the roles played by NGOs, there are other opportunities for public participation in the CITES COP. These avenues have a significant, yet largely untapped, potential for increasing procedural regard for affected states' interests. For example, there are participation mechanisms allowing rural communities impacted by CITES listings to voice their views as part of listing debates, although these opportunities are used infrequently. There has been increased awareness of,

106 TRAFFIC, 'Traffic and CITES' (2017) <www.traffic.org/cites/>.

107 See, eg, D Bodansky, 'Legitimacy' in D Bodansky, J Brunnée and E Hey (eds), *The Oxford Handbook of International Environmental Law* (Oxford University Press, 2007) 704, 718; D C Esty, 'Good Governance at the Supranational Level: Globalizing Administrative Law' (2006) 115 *Yale Law Journal* 1490, 1531–1532.

108 See generally Duffy, 'Global Environmental Governance and North–South Dynamics', above n 7.

109 *Ibid* 231; see also Couzens, *Whales and Elephants*, above n 12, 135–136.

110 IUCN/TRAFFIC, 'IUCN/TRAFFIC Analyses of the Proposals to Amend the Appendices for CITES CoP16' (2013) <https://cmsdata.iucn.org/downloads/iucn_traffic_analyses_of_proposals_cites_cop16_2012_1.pdf>; Montazeri, 'Protecting the Pachyderm', above n 4, 137–141.

and sensitivity to, the impact of listing decisions on rural communities' livelihoods in CITES since the 1990s and 2000s, which reached a high point at COPs 16 and 17 in 2013 and 2016, respectively.[111] This shift has created a more favorable environment for participation of representatives from rural communities to enhance decision makers' regard for their interests in listing decisions, which, as previously noted, directly alter the grounds for non-compliance.

While there is currently no requirement for rural communities' livelihood interests to be taken into account in listing decisions, such considerations are frequently discussed as part of listing debates in practice. The participation of representatives of rural communities can powerfully impact these debates.[112] For example, Cooney and Abensperg-Traun note that the participation of representatives of the Inuit community,[113] who described the likely cultural and financial impacts that an Appendix I listing for polar bears would have on them, had a 'strong influence' on the debate and the final COP vote that rejected the proposal to up-list polar bears at COP 16.[114] These authors argue that '[h]earing directly from people culturally and economically linked to the species in question about their views on a decision – how it would affect them and how it would affect conservation – carries considerable weight'.[115] Thus, avenues for public participation have demonstrated, if as yet under-utilized, potential for increasing procedural regard for affected communities' interests in decisions shaping the parameters of non-compliance.

2.3.2 Decision makers and decision-making processes in the Standing Committee

There are juridical participation rights afforded to Parties that are subject to compliance proceedings before the Standing Committee, which prima facie aligns with procedural regard. Once the Standing Committee has decided to proceed with a compliance matter, the Party concerned is given 'the opportunity

111 For example, at COP 16, a resolution on 'CITES and Livelihoods' was adopted, which was subsequently revised at COP 17: CITES, CITES and Livelihoods, Resolution Conf 16.6 (Rev COP 17), Oct 2016. This resolution recognizes that the implementation success of CITES is likely to be enhanced with the engagement of rural communities that traditionally derive their livelihoods from CITES-listed species. One NGO described COP 17 (Johannesburg, 2016) as the COP where 'Africa and rural communities have had their biggest voice': IISD Reporting Services, 'Summary of the Seventeenth Meeting of the Conference of the Parties to the Convention on International Trade in Endangered Species of Wild Fauna and Flora' (8 October 2016) 21(97) *Earth Negotiations Bulletin* 22 <www.iisd.ca/download/pdf/enb2197e.pdf>.

112 Cooney and Abensperg-Traun, 'Raising Local Community Voices', above n 13, 303, 305–307, 309.

113 The Inuit representatives at COP 16 participated both as Inuit organizations with observer status, and as part of the Canadian delegation: *ibid* 309.

114 *Ibid* 305–306, 309. Other salient factors included the current status of polar bears and the likely impacts of climate change: *ibid* 306.

115 *Ibid* 309.

to provide comments within a reasonable time limit'.[116] The Party concerned also has the right to participate in discussions with respect to its own non-compliance proceedings, and may request guidance regarding possibilities for financial assistance from the Secretariat or the Standing Committee for this purpose.[117] Hence, some features of administrative decision making pertaining to the right to be heard are evident in compliance proceedings before the Standing Committee. However, as previously noted, political considerations may influence which compliance matters even get to this stage.

Reforms to membership rules for the Standing Committee have aimed to decrease the risk of disregard for developing states' interests in compliance decision making. However, despite this, committee membership appears to work in favor of states that are showing signs of non-compliance. The Standing Committee consists of 16 representatives from member countries, who are elected by the COP[118] to achieve balanced regional representation. Prior to 1994, the regional members from North America and Oceania had disproportionate voting influence vis-à-vis other regional members.[119] Reforms of the Standing Committee's membership rules at COP 9 (Fort Lauderdale, 1994) facilitated proportional representation, which signified greater decisional participation rights for less powerful states.[120] However, not once since 1985 has any Standing Committee member country had a trade sanction imposed against it.[121] This is despite the fact that, as discussed in relation to Japan above,[122] there are situations in which the country of origin of Standing Committee members appears to have contravened requirements that ordinarily trigger trade suspensions. This suggests that compliance is negotiable, particularly for Parties with representatives on the Standing Committee, which cuts strongly against the grain of administrative decision-making ideals.

2.3.3 Decision-making processes in the COP

While COP decision making is primarily political, steps have been taken to overcome the practical obstacles to developing countries' participation in COP

116 CITES compliance procedures, above n 3, para 25.

117 *Ibid* paras 27 and 28.

118 CITES, Establishment of Committees, Resolution Conf 11.1 (Rev COP 17), Oct 2016, Annex 1, (a)(III). In addition, the Committee has representatives from Switzerland, as the depositary government, and the previous and next host countries, thus bringing the total committee membership to nineteen.

119 J M Hutton and B Dickson, *Endangered Species, Threatened Convention: The Past, Present and Future of CITES, the Convention on International Trade in Endangered Species of Wild Fauna and Flora* (Earthscan, 2000) 117.

120 CITES, Regional Representation on the Standing Committee, document prepared by Malawi on behalf of the African region for COP 9, Doc 9.7, Nov 1994; CITES, Establishment of Committees, above n 118, Annex 1.

121 In only two instances, concerning Vanuatu in 2002 and India in 2004, has a trade embargo been recommended for an alternate member: Sand, 'Enforcing CITES', above n 20, 263.

122 See Chapter 6, section 2.1.2.

meetings, facilitating opportunities for procedural regard. COP decisions are taken by a two-thirds majority of Parties present and voting,[123] which is a higher threshold than the simple majority vote required for Standing Committee decisions. To facilitate participation at COP meetings by representatives from least developed and developing countries, in 1986 a Sponsored Delegates Project was developed. This is significant, as developing countries' participation is vital for promoting regard for their interests, but in many intergovernmental decision-making forums is hampered by a lack of funding to meet travel and other expenses associated with attending meetings.[124] At recent COPs (1987–2013), an average of 160 delegates from 100 countries were sponsored under this project.[125] The Sponsored Delegates Project is a valuable mechanism for ensuring that developing states have opportunities to voice their views. This mechanism supports informed and balanced decision making, and therefore the aims of procedural regard.

2.4 Reason giving in the CITES compliance system

The CITES compliance system has a poor record in relation to reason giving. The types of transparency issues outlined above are interlinked with the relative dearth of reason giving in the CITES compliance system. The Secretariat's move away from detailed, comprehensive non-compliance reporting in infractions reports toward more limited and piecemeal compliance reporting since COP 11 in 2000 has reduced the extent of reason giving informing the initiation of compliance proceedings. Moreover, the regular use of secret ballot voting by the Standing Committee and COP obscures the bases upon which decisions are adopted or rejected, including whether decisions are being taken in accordance with agreed criteria specified in the Convention.[126]

The reason-giving requirements in the CITES compliance procedures reflect an intention for decision makers to consider less powerful states' interests in compliance determinations. There is little evidence, however, that this occurs in practice. Paragraph 32 of these procedures lists four considerations that must be clearly set out in the Standing Committee's recommendations:

a) The capacity of the Party concerned, especially developing countries, and in particular the least developed and small island developing States and Parties with economies in transition;

123 COP Rules of Procedure, above n 56, rule 26, para 1.

124 See, eg, B S Chimni, 'Co-option and Resistance: Two Faces of Global Administrative Law' (2005) 37 *International Law and Politics* 799, 813–814.

125 CITES, Sponsored Delegates Project, document prepared by the European Union and Senegal for COP 17, Doc 8, Jun 2016. Similarly, the Secretariat has been instructed by the Standing Committee to make provision in its budget for the payment, if requested, of 'reasonable and justifiable travel expenses' for one member from each region to attend Standing Committee meetings: CITES, Establishment of Committees, above n 118.

126 CITES, Proposal to Improve Transparency of Voting during Meetings of the COP, above n 90, para 8.

b) Such factors as the cause, type, degree and frequency of the compliance matters;
c) The appropriateness of the measures so that they are commensurate with the gravity of the compliance matter; and
d) The possible impact on conservation and sustainable use with a view to avoiding negative results.

These reason-giving requirements appear to be designed to promote procedural regard for the interests of developing states. However, in the publicly available Summary Records of the Standing Committee meetings available on the CITES website, which include the Committee's recommendations, there is a notable absence of any such reasons since the compliance procedures were adopted in 2007.[127]

Indeed, a document prepared by the Secretariat for COP 17 (Johannesburg, 2016) acknowledges that the causes of non-compliance should be taken into account by the Standing Committee, but are not always 'explicitly set out in Standing Committee's recommendations' as required by the compliance procedures.[128] The Secretariat recommended that the COP should direct the Standing Committee to assess the factors and criteria to be taken into account when deciding on compliance measures in accordance with the CITES compliance procedures.[129] However, the European Union, supported by the United States, suggested deleting this text.[130] This deletion occurred before the other recommendations for improving compliance were agreed to by the COP.[131] Hence, while the CITES compliance procedures invite greater consideration of less powerful states' circumstances and interests in compliance decisions, to date the Parties have not prioritized this opportunity for demonstrating procedural regard.

2.5 Review in the CITES compliance system

Avenues for review and appeal in the CITES compliance system are limited insofar as there are no formal avenues for states to appeal to the COP regarding the Standing Committee's compliance recommendations. As previously mentioned, review can also encompass a body that is higher in the compliance hierarchy exercising oversight over a subsidiary body's decision making in the absence of a formal appeal.[132] In this regard, the CITES compliance procedures stipulate that decisions of the Standing

127 See the Summary Records for Standing Committee meetings for the fifty-sixth to sixty-ninth meetings available at: CITES, 'The Standing Committee', above n 32.
128 CITES, CITES Compliance Matters, prepared by the Secretariat for COP 17, above n 38, para 33.
129 Ibid Annex 1, para 17.B(b).
130 IISD Reporting Services, 'Summary of the Seventeenth COP', above n 111, 6.
131 CITES, 'Draft Decisions on Legal Acquisition Findings and Compliance Matters', COP 17, Com. II. 13, Oct 2016, para 17.B.
132 See Chapter 2, section 3.5.

Committee pertaining to compliance are subject to review by the COP 'as needed'.[133] In practice, it is common for Standing Committee decisions recommending a trade suspension in CITES-listed species to be communicated to the Parties by the CITES Secretariat without the requirement for official endorsement by the next COP, which may be up to three years away.[134] The COP's role in reviewing the Standing Committee's compliance determinations does indeed seem to occur on a needs basis, rather than as routine oversight of the extensive powers delegated to the Standing Committee. On balance, the general dearth of avenues for appeal or review in the CITES compliance system limits opportunities for procedural constraints and regard.

3 A vignette of compliance decision making: the United Arab Emirates (UAE)

The United Arab Emirates (UAE) was subject to recommended trade suspensions for non-compliance in the mid-1980s and early 2000s, and is thus one of the few Parties that has been subject to trade sanctions on multiple occasions. The compliance proceedings concerning the UAE at these two different historical junctures highlight an increasing reliance on administrative procedures over time, accompanied by greater potential for constraints upon politics and procedural regard. Although the second non-compliance proceedings concerning the UAE fail to satisfy due-process ideals in IEL,[135] they nonetheless illustrate the emergence over time of an environment that is more conducive to procedural regard.

3.1 The 1985 non-compliance proceedings

The 1985 non-compliance proceedings concerning the UAE were largely political in nature, and weakly procedurally constrained. These proceedings were triggered after the Secretariat received extensive information indicating that illegal trade in a number of specimens of CITES species, particularly elephant ivory, rhinoceros horn and snake skins, was 'occurring freely' in and through the UAE.[136] After multiple attempts at facilitative dialogue from the Secretariat were ignored or unsuccessful, the UAE communicated to the Secretariat that it was yet to take domestic measures to implement CITES. Consequently, the Standing Committee discussed the matter in November 1985,[137] and the Secretariat issued Notification No. 366 urging all Parties to prohibit trade in CITES-listed species with the UAE. This was one of the

133 CITES compliance procedures, above n 3, para 10(c).
134 Davies, 'Non-Compliance', above n 54, 99–100.
135 A closer approximation to these ideals can be found in the design of the Kyoto compliance system's Enforcement Branch: see Chapter 5.
136 CITES, Review of Alleged Infractions, prepared by the Secretariat for COP 6, Doc 6.19 (Rev), Jul 1987, 560–561.
137 *Ibid.*

earliest uses of the trade-suspension recommendation by the Standing Committee. Although the UAE was held to account for its CITES commitments via this process, this was essentially a political process with little evidence of administrative mechanisms for transparency, participation, reason giving and review.

Thereafter, the Secretariat continued to attempt to engage in facilitative dialogue with the UAE to promote the implementation of CITES controls in that country, including by sending a Secretariat staff member to the UAE in November 1986. However, the UAE government declined to meet with this staff member, and sent a telex to the Secretariat informing it of the President's decree to withdraw from the Convention.[138] The Secretariat's requests for bilateral discussions of this decision were denied, and on 27 January 1988, the UAE's withdrawal from CITES became effective.[139] This highlights one of the risks associated with MEA compliance systems taking strong action on non-compliance issues – that sovereign states will elect to exit the Convention, and thus operate outside the treaty's purview. In response to the UAE's withdrawal, the Secretariat issued a notification to Parties in March 1987 reinforcing 'a total ban on all trade in CITES species' with the UAE, which is permissible under provisions in Article X designed to ensure that trade between Parties and non-Parties does not undermine the aims of the Convention.

In 1988–1989, CITES Parties used diplomatic channels to address illegal trade in ivory, which was flourishing in the UAE after it renounced treaty membership, and to encourage the UAE to consider rejoining CITES. While originally a high-level delegation, comprising a Standing Committee representative from one African and one Arabic country, and from the United States as the Committee chair, was proposed, ultimately a 'quieter approach at the ambassadorial level' was favored.[140] Subsequent contacts by ambassadors from the US, the UK and the Netherlands were instrumental in catalyzing the closure of two ivory-carving factories in Dubai, and paving the way for the UAE to rejoin CITES.[141] When the UAE did rejoin the Convention in February 1990 and introduced new implementing legislation, the recommended trade ban was lifted.[142] This was a successful utilization of diplomatic techniques to promote treaty and thereby compliance aims. It also points to the potential limits of administrative decision making in coping with high-level politics in IEL if non-compliance issues escalate.

3.2 The 2001 non-compliance proceedings

In contrast, the second recommended trade suspension concerning the UAE in 2001 evidences an increasing reliance on administrative procedures and decision making, alongside facilitative practices. After the UAE rejoined the Convention, its

138 *Ibid* 561.
139 *Ibid.*
140 Reeve, *Policing International Trade in Endangered Species*, above n 20, 105.
141 *Ibid*, citing CITES, Summary Report, SC 18, Mar 1989, 4.
142 Sand, 'Whither CITES?', above n 5, 38–39.

compliance record was not in the spotlight again[143] until the Secretariat's infractions report for COP 11 (Gigiri, 2000) drew attention to a formal communication it had made to the UAE regarding non-compliance with Article XIII(1) of the Convention. This communication was triggered by evidence that false or forged Russian Federation documents were being used to obtain re-export certificates for caviar shipments.[144] Thereafter, at the 45th Standing Committee meeting in June 2001, the Secretariat drew the Committee's attention to its 'increasing concern regarding illegal trade', including 'smuggling of CITES-listed specimens', such as falcons and cheetah cubs, to the UAE.[145] The Secretariat identified 'regularly receive[d] reports, from a variety of official and informal sources'[146] as being the basis for their concerns. Presumably, some of this performance-review information was provided by the NGOs that frequently participate in CITES review processes, but the reference to unidentified 'informal sources' raises questions about the Secretariat's adherence to transparency ideals and formal participation channels.

The Secretariat's proposed response to the UAE's alleged infractions was reviewed by the Standing Committee. The Secretariat noted that, as part of its response to these alleged infractions, it had offered to conduct a mission to the UAE to discuss enforcement issues in detail, but had received no response to this invitation.[147] Notably, the Secretariat stopped short of recommending a second trade suspension. The Standing Committee acted autonomously in deciding that, if the UAE did not invite such a mission allowing the Secretariat to satisfy itself that the Convention was being adequately implemented by 31 October 2001, the Secretariat would distribute a notification to the Parties recommending a trade suspension in CITES-listed species.[148] That is, after reviewing the Secretariat's work, the Standing Committee was willing to hold the UAE to a higher standard of accountability than the Secretariat originally proposed. In the light of the UAE's previous rescission of CITES, the Standing Committee's stance on this issue was courageous, evidencing a prioritizing of accountability aims over political concerns about treaty exit.

The subsequent developments reflect elements of both administrative decision-making and facilitative practices. Immediately before the deadline established by the Standing Committee, the Secretariat received an invitation from the UAE, and a mission was undertaken in late October and early November 2001.[149] In accordance with accountability, transparency and participation aims, the Secretariat

143 This was evidenced by the lack of attention given to the UAE in the Secretariat's infraction reports for COP 9 and COP 10: see CITES, Review of Alleged Infractions for COP 9, above n 74; and CITES, Review of Alleged Infractions for COP 10, above n 75.

144 See Decision 10.122 reported in CITES, Review of Alleged Infractions for COP 11, above n 71, para 48.

145 CITES, Enforcement Matters, prepared by the Secretariat for the 45th Standing Committee meeting, SC45 Doc 11.2, Jun 2001, paras 10 and 11.

146 *Ibid* para 11.

147 CITES, Enforcement Matters (SC 45), above n 145, para 13.

148 CITES, Summary Report, prepared for the 45th Standing Committee meeting, SC45, Jun 2001, 9–10.

149 CITES, Enforcement Matters, prepared by the Secretariat for the 46th Standing Committee meeting, SC46 Doc 11.2, Mar 2002, paras 10 and 11.

met with 'a range of government officials and non-government consultants engaged in environmental and wildlife issues' and Management Authority staff[150] in order to accurately gauge the extent and causes of the UAE's non-compliance. On the basis of this extensive information-gathering exercise, the Secretariat concluded that the UAE was still not adequately enforcing the Convention. Accordingly, the Secretariat issued a notification implementing the Standing Committee's previous trade-suspension recommendation.[151] However, even after taking this strong enforcement action, the Secretariat maintained contact with the CITES Management Authority of the UAE and continued to provide practical and technical assistance,[152] indicating concomitant reliance on facilitative techniques that take into account country-specific interests.

This second trade suspension proved to be effective in significantly enhancing the UAE's compliance record. At the 46th Standing Committee meeting (Geneva, 2002), the Secretariat noted that the UAE had 'responded positively' to the trade recommendation. Specifically, it had ceased to authorize trade in CITES specimens, and developed an 'excellent' new federal law to implement CITES.[153] The trade suspension was withdrawn in phases[154] and, in November 2002, the Standing Committee authorized the Secretariat to issue a notification completely withdrawing the trade suspension.[155]

Therefore, compared to the 1985 proceedings, the second non-compliance proceedings concerning the UAE demonstrate greater reliance on select administrative procedures, and concurrent utilization of facilitative practices to achieve compliance aims. This underscores the gradual but notable proceduralization in the CITES compliance system from the mid-1980s to the early 2000s, which placed some constraints upon traditional political approaches to compliance decision making, and offered greater opportunities for regard for affected states' interests. Although the effects of adopting administrative practices for accountability, transparency and participation fall short of proceduralization ideals, the foundations for procedural constraints on power politics and greater regard for affected states' interests are evident in the second non-compliance proceedings concerning the UAE in the early 2000s.

4 Conclusion

Over multiple decades of practice, increasing proceduralization has been evident in the CITES compliance system. This trend was reinforced by the release of the

150 *Ibid* para 11.
151 *Ibid* Notification to the Parties No. 2001/079 on 20 November 2001.
152 *Ibid* para 13.
153 CITES, Summary Report, prepared for the 46th Standing Committee meeting, SC46, Mar 2002, 12–13.
154 CITES, Enforcement Matters, prepared by the Secretariat for the 47th Standing Committee meeting, SC47 Doc 10.2, Nov 2002, para 4.
155 CITES, Summary Report, prepared for the 47th Standing Committee meeting, SC 47, Nov 2002. There is no information provided regarding the content of these interventions.

Guide to CITES Compliance Procedures in 2007, although these procedures remain non-binding. In particular, there has been adoption of administrative procedures and mechanisms pertaining to accountability and participation. The system's record in relation to transparency and reason giving is non-linear, and demonstrates both gains and losses. Avenues for review are largely non-existent. Despite this variable record, when viewed holistically, the compliance system is undoubtedly more proceduralized today than it was at its inception in the 1980s.

The CITES compliance system's patchy proceduralization is reflected in the mixed success of administrative procedures in creating procedurally constrained zones for decision making. A notable example of administrative procedures serving to constrain political influences within the compliance system is the accountability mechanisms that have curbed the historical excesses of the Secretariat. However, at around the same time as these accountability gains, the compliance system's previously strong record in transparent infractions reporting was eroded by a shift toward less extensive and informative publicly available compliance information. Moreover, despite an ostensible commitment to administrative decision-making ideals such as transparency, political practices such as secret ballots continue to prevail. Thus, while the adoption of administrative procedures is associated with increasingly procedurally constrained decision making in some domains, the failure to adopt new, or the loosening of existing, procedural controls in other domains has meant that some zones of decision making remain relatively unfettered. This helps to account for the mix of administrative and diplomatic decision-making styles evident in this compliance system.

Administrative procedures' potential to promote procedural regard for affected states' interests does not have a particularly strong record within the CITES compliance system. The Standing Committee's failure to adhere to the compliance procedure that requires reason giving informed by norms of differential treatment is a missed opportunity for demonstrating procedural regard. Furthermore, the Standing Committee's apparent bias against developing states in compliance decisions is an area of significant concern in terms of both procedural and substantive disregard. The risk that without adequate due-process guarantees, compliance decisions will be strongly shaped by differences in power between states appears to be borne out by the CITES experience, reinforcing the importance of robust procedural safeguards to counterbalance this risk.

Nonetheless, there are lessons that can be learned from this case study regarding favorable conditions for procedural regard. Since the 1990s, there has been a shift away from a narrow preservationist agenda dominated by powerful Northern states and NGOs toward a greater recognition of the link between sustainable development and species conservation for some developing range states. This paradigm shift has created a context of pluralism and contestation that is more conducive to procedural regard for affected states' interests and preferences in listing decisions and non-compliance proceedings. Participation by NGOs and representatives of communities impacted by CITES listings has also enhanced regard for affected states' interests. Such developments indicate potential for realizing the benefits of proceduralization, even if current practices fall short of this ideal.

7 The benefits of administrative procedures in MEA compliance systems

1 Introduction

The Montreal Protocol, Kyoto Protocol and CITES address critical global environmental issues of ozone-layer depletion, climate change and trade in endangered species, respectively. This study has developed a framework for evaluating the desirability of adopting administrative procedures, and applied it to the compliance systems of these three major global MEAs. In Chapter 1, I argued that proceduralization is important in these contexts to (1) demonstrate respect for states' autonomy, and diverse capacities and circumstances, (2) understand the cause of non-compliance so that appropriately tailored and responsive compliance solutions can be adopted and (3) justify the imposition of potentially intrusive consequences for non-compliance. However, there is a risk that these benefits remain unrealized if administrative procedures are co-opted by, and reinforce the interests of, the powerful.[1] Accordingly, this book has analyzed the roles played by administrative procedures using evaluative criteria linked to concrete, institutional practices. This analysis shows that, under certain conditions, the adoption of administrative procedures limits the influence of asymmetric power relations on compliance decision making, and enhances regard for the interests of affected states, which are typically developing and transition countries. This is significant, as it shows that administrative procedures can help to level the playing field for less powerful states by increasing the opportunities for their compliance interests to be voiced and considered in MEA decision-making processes.

This overarching conclusion is informed by three main findings. First, although there are notable differences between the extent and implications of proceduralization in each case study, there is a general trend toward increasing adoption of administrative procedures. Second, in some domains, these procedures have contributed to creating procedurally constrained spaces for compliance decision making, such as in the Kyoto Compliance Committee. However, the net effect of

1 See, eg, B S Chimni, 'Co-option and Resistance: Two Faces of Global Administrative Law' (2005) 37 *International Law and Politics* 799, 803.

the power-constraining qualities of administrative procedures is frequently undermined by the development of zones of compliance decision making with thinner procedural safeguards, and thus fewer constraints on diplomatic and negotiation-based practices. Third, there is little evidence of administrative procedures being co-opted by powerful states. Administrative procedures have demonstrated potential to redress existing power imbalances by bringing to the fore less powerful states' interests, with the Montreal compliance system providing an exemplar in this regard. States' ability to use procedural mechanisms plays an important role in realizing latent procedural regard opportunities.

In addition to these aggregate findings, a number of context-specific trends regarding favorable conditions for procedural regard can be identified. Specifically, the Montreal compliance system case study highlights that mechanisms for procedural regard can be valuably reinforced by conditional funding to support Parties' return to compliance, promoting greater engagement with existing opportunities for interest representation. The Kyoto compliance system yields valuable lessons regarding the importance of facilitating the use of procedurally constrained decision-making spaces where they are available to prevent a shift to more informal negotiating spaces that lack due-process guarantees. The CITES compliance system case study shows that pluralism within the regulatory discourses animating a regime can create a context that is more supportive of procedural regard for affected states' interests and preferences than one that is premised on an ostensibly neutral consensus between states. This case study also illustrates the potential benefits of participation by civil-society representatives who can expand the range of perspectives factored into decisions that affect compliance, including the interests of affected local communities.

The next section elaborates on this brief overview of the findings from the case study analysis. Section 3 then reflects upon the lessons learned for future compliance processes in international environmental law (IEL). Specifically, an analysis of the emerging compliance apparatus under the Paris Agreement underscores that the proceduralization framework provides an evaluative tool, as well as a practical guide for the future development of MEA compliance frameworks. The chapter concludes with remarks reinforcing that, despite some implementation challenges, adopting administrative procedures is a desirable focal point for the design of MEA compliance systems.

2 A synthesis of the case study findings

2.1 The extent of adoption of administrative procedures

Although the configuration and implications of administrative procedures varies in each of the three case studies, there is a clear trajectory toward greater proceduralization in the selected MEA compliance systems. Of the three, the Kyoto and CITES compliance systems represent the highest and lowest levels of proceduralization, respectively. The Kyoto compliance system, which was developed later than its counterparts in CITES and the Montreal Protocol, represents an administrative apparatus that from the outset incorporated extensive procedures for accountability, transparency, participation, reason giving and review. It thus represents the high water mark of reliance on administrative procedures in global MEA compliance systems.[2]

In both the Montreal and CITES compliance systems, there has been a subtle but notable shift toward greater reliance on administrative procedures over time. In the Montreal compliance system, avenues for review are circumscribed but there is evidence of the adoption of a limited set of administrative procedures for accountability, transparency, participation and reason giving. In the CITES compliance system, proceduralization has been particularly evident in relation to clarifying and tightening accountability expectations, and increasing avenues for participation. In contrast, transparency and reason giving are two areas in which little progress has been made, and in the case of transparency, previous gains have been eroded over time. Despite this, on balance, the overall picture that emerges from the case studies is one of increasing reliance on administrative procedures. As is explored further in the following sub-sections, this phenomenon has varying significance in each context.

More evidence is required to assess whether the trend toward adoption of administrative procedures in these three compliance systems is occurring in other global MEAs. As discussed in Chapter 3, MEAs differ with respect to, inter alia, the intrusiveness of the potential consequences for non-compliance, who holds decision-making authority, the degree of interdependence required to achieve collective goals and the political sensitivity of the issues that they address.[3] In the light of this contextual variation, further research is required to establish the extent to which the trends observed in the case-study MEAs are indicative of broader trends in global environmental governance.

2.2 The role of administrative procedures in creating procedurally constrained spaces for decision making

The case study analysis highlights that administrative procedures play important roles in creating spaces for decision making in which political influences on compliance decisions are constrained. However, frequently these gains are offset by the prevalence and re-assertion of decision making in other domains of compliance decision making with less stringent procedural safeguards. This trend underscores the desirability of more comprehensive proceduralization to minimize pockets of relatively unfettered political decision-making.

In both the CITES and Kyoto compliance systems, increasing reliance on administrative procedures has resulted in procedurally constrained decision making

2 See also J Gupta, 'Developing Countries: Trapped in the Web of Sustainable Development Governance' in O Dilling, M Herberg and G Winter (eds), *Transnational Administrative Rule-Making: Performance, Legal Effects, and Legitimacy* (Hart Publishing, 2011) 305, 309.

3 D C Esty, 'Good Governance at the Supranational Level: Globalizing Administrative Law' (2006) 115 *Yale Law Journal* 1490, 1509–1514.

in some, but not all, domains. For example, within the CITES compliance system, the Standing Committee's use of accountability mechanisms such as independent inspections of internal operations and the imposition of budgetary controls contributed to disciplining the excesses of the Secretariat in the 1990s.[4] However, as the accountability of the Secretariat's compliance operations improved in the late 1990s and early 2000s, previous transparency gains in relation to the Secretariat's infractions reporting – and hence publicly available compliance information – were eroded. Thus, as procedural constraints were introduced in one area of compliance decision making, the procedural controls in another area were loosened, making it difficult to gauge the overall net gains arising from these changes.

The Kyoto compliance system provides evidence of a migration of regulatory power away from the procedurally constrained Compliance Committee to ERTs, whose processes are relatively informal. At first glance, this compliance system is exceptional in terms of its attempts to create zones for decision making in both the Facilitative and Enforcement Branches that are circumscribed by extensive due-process guarantees. Despite these laudable efforts, the analysis in Chapter 5 demonstrates a concentration of decision-making authority in the ERTs. These small teams of experts have served as unofficial gatekeepers for the compliance system, and yet their processes are relatively opaque. The bottleneck created by the ERTs has curtailed the number of cases heard before the Enforcement Branch, and the facilitative roles played by these small teams effectively rendered obsolete the Facilitative Branch.[5] The compliance rules and procedures point to a predominantly technical role for ERTs, and a more extensive, yet procedurally constrained, facilitative role for the aptly named Facilitative Branch. The ERTs' significant facilitative roles in practice underscore the risk of decision-making authority gravitating toward less procedurally constrained spaces, hampering the realization of desirable aims of proceduralization.

Similarly, the benefits stemming from adopting administrative procedures in the Montreal compliance system are uneven. In some areas, the increasing use of accountability and reason-giving procedures has promoted consistent treatment of comparable instances of non-compliance, with leeway to take into account country-specific factors. In other domains, however, adopting procedures to place checks on decision making, such as those pertaining to expert inputs, participation by civil-society representatives and review, has not been prioritized. Furthermore, compliance standards that are treated as negotiable and revisable raise quality

4 Applying insights from international relations scholarship, this may be seen as an example of states as principal actors attempting to achieve control over an international bureaucracy after authority has been delegated: see, eg, D G Hawkins, D A Lake, D L Nielson and M J Tierney, 'Delegation under Anarchy: States, International Organizations and Principal-Agent Theory' in D G Hawkins, D A Lake, D L Nielson and M J Tierney (eds), *Delegation and Agency in International Organizations* (Cambridge University Press, 2006) 3, 30–31.

5 A Zahar, *International Climate Change Law and State Compliance* (Routledge, 2015) Chapter 3.

concerns and, in some cases, undermine strict adherence to binding legal rules. Thus, again, patchy proceduralization detracts from the power-constraining qualities of administrative procedures utilized in other parts of the compliance system.

In sum, the case-study analysis shows that administrative procedures do play important roles in creating procedurally constrained zones for compliance decision making. However, the benefits of increasing procedural safeguards in some domains are frequently undermined by the development of, and migration of regulatory power to, other pockets of compliance decision making in which such safeguards are less developed. In these pockets, there continues to be a prevalence of diplomatic, co-operative and negotiation-based modes of compliance decision making that lack extensive due-process guarantees. This finding is consistent with Kingsbury's observation that adopting '[r]ules of procedure in a deliberative body ... may encourage defection by powerful interests to make decisions in a more amenable forum'.[6] It is perhaps because administrative procedures are effective in constraining political influences that such defection is apt to occur.

On a practical level, the emergence of spaces with fewer constraints on political decision making is not entirely unexpected in a context in which states are the creators and enforcers, as well as the agents and subjects, of compliance rules and procedures.[7] In normative perspective, however, this tendency underscores the benefits of comprehensive, system-wide proceduralization to limit the opportunities for compliance decision making to occur in less procedurally constrained spaces. This is because relatively unfettered diplomatic and negotiation-based practices risk disadvantaging states with less bargaining power, which is morally and practically undesirable. These findings reinforce the value of examining the roles played by administrative procedures within a broader regulatory framework, such as MEA compliance systems, which allows these types of systemic tendencies regarding the movements of regulatory power to be brought into relief, aiding identification of priorities for future procedural design.

2.3 The role of administrative procedures in promoting procedural regard for affected states' interests

The case study analysis also shows that administrative procedures play constructive roles in promoting procedural regard for affected states' compliance

6 B Kingsbury, 'The Concept of "Law" in Global Administrative Law' (2009) 20(1) *European Journal of International Law* 23, 35. Benvenisti makes a similar point more starkly: '[T]he application of legal constraints to decision-makers is never a simple exercise, because those who cherish unfettered discretion are tempted to manipulate the new rules and create new islands of arbitrariness and impunity': E Benvenisti, *The Law of Global Governance* (Hague Academy of International Law, 2014) 287.

7 B Kingsbury, N Krisch and R B Stewart, 'The Emergence of Global Administrative Law' (2005) 68 *Law and Contemporary Problems* 15, 36; J Waldron, 'Are Sovereigns Entitled to the Benefit of the International Rule of Law?' (2011) 22 *European Journal of International Law* 315, 327–337.

interests.[8] In the CITES and Kyoto compliance systems, the potential of administrative procedures to facilitate consideration of less powerful states' interests has tended to remain latent, for varying reasons in each instance. Nonetheless, both case studies provide valuable insights into conditions that are conducive to procedural regard. The Montreal compliance system exemplifies the use of administrative procedures in a way that promotes regard for affected states' interests, and highlights the desirability of conditional funding to support states' compliance interests that have been elicited through the non-compliance process.

The CITES compliance system exemplifies the risk that, without adequate procedural safeguards, compliance processes will be strongly influenced by differences in power between states. In this system, where pre-existing problems of disregard for weaker states' interests existed, the introduction of select administrative procedures over time neither exacerbated nor significantly ameliorated the problem, but rather appeared to perpetuate pre-existing power relations and asymmetries. There are significant missed opportunities for procedural regard in this compliance system, including a failure to utilize the compliance procedure that requires reason giving informed by norms of differential treatment. Of even greater concern is the apparent bias against developing states in the Standing Committee's accountability processes underpinning its trade-suspension recommendations. Couzens aptly describes CITES as an 'old watchdog', and there is much to be learned from the loopholes and weaknesses that have been exploited by states in these older-style treaties.[9]

Despite these sobering findings, it is possible to distill lessons from the CITES compliance experience regarding conditions that enhance regard. In the CITES compliance system, the growing traction of sustainable-use discourses alongside the traditional preservationist agenda has created an environment that is more receptive to procedural regard for affected states' diverse interests and preferences in listing decisions and non-compliance proceedings. In relation to the former, participation by NGOs and representatives of rural communities impacted by CITES listings has demonstrated potential to enhance regard for range states' livelihood interests in sustainable use in decisions that determine the parameters for compliance within the regime. Therefore, although there are ongoing and significant problems of procedural and substantive disregard for less powerful states' interests within the CITES compliance system, there is some evidence of positive shifts toward increasing opportunities to give voice to affected interests.

The Kyoto compliance system has considerable potential for promoting procedural regard, which has only been partially realized in practice. In this

8 Cf Prosper Weil, who criticizes softer legal arrangements, and argues that it is 'hard law with its rigor … that comes between the weak and the mighty to protect and deliver': P Weil, 'Towards Relative Normativity in International Law?' (1983) 88 *American Journal of International Law* 413, 442.

9 E Couzens, *Whales and Elephants in International Conservation Law and Politics: A Comparative Study* (Routledge, 2014) 6, 121.

system, diminished access to the Facilitative and Enforcement Branches caused by the ERT bottleneck has limited opportunities for states to benefit from the extensive due-process guarantees offered in these branches. The institutional designers of this system appeared to envisage that all states' capacity limitations and compliance interests could be taken into account in the Facilitative Branch's processes, while industrialized states would be held to stricter standards of accountability by the Enforcement Branch. However, while the Facilitative Branch holds promise for demonstrating regard for affected states' compliance interests in a forum that is constrained by procedural safeguards, this potential remains effectively untapped. The Enforcement Branch adopts an automatic and non-discretionary approach to responding to non-compliance, which accords with the accountability aim of like treatment of analogous cases, but does not reflect the flexibility to consider country-specific circumstances that is characteristic of procedural regard. Due to their opaque and informal processes, the extent to which ERTs have provided procedural regard for the interests of affected states is unclear, and the relative dearth of procedural safeguards shaping ERTs' work draws into question the consistency of any such processes. This underscores the importance of facilitating states' access to available procedural regard opportunities, such as those available through the Facilitative Branch.

Of the three case studies, the Montreal compliance system best exemplifies the use of administrative procedures to promote procedural regard for affected states' interests. In this system, the compliance interests of developing states and CEITs have increasingly been brought to the fore as greater utilization of administrative procedures occurred over time. An overriding aim of the procedures developed appears to be maximizing access to information to understand Parties' non-compliance, and supporting their return to compliance through flexible yet proceduralized approaches, reinforced by conditional funding. The possibility of this funding support appears to play an important role in buttressing the procedural regard opportunities available, and provides a mechanism for addressing some of the compliance needs and interests articulated by affected states. In isolation, the provision of financial and technical assistance to address the causes of non-compliance is an example of substantive regard, in which decision-making outcomes are favorable to less powerful actors.[10] Importantly, however, such responsiveness is made possible and justifiable by the dialogue between states and other actors that is facilitated by the extensive procedural regard opportunities in this compliance system. Although the specific substantive provisions of the Montreal Protocol are likely to be politically difficult to replicate,[11] its compliance system provides valuable lessons regarding efficacious procedural design.

10 R B Stewart, 'Remedying Disregard in Global Regulatory Governance: Accountability, Participation, and Responsiveness' (2014) 108 *American Journal of International Law* 211, 214.

11 See generally C R Sunstein, 'Of Montreal and Kyoto: A Tale of Two Protocols' (2007) 31(1) *Harvard Environmental Law Review* 1; E DeSombre, 'The Experience of the Montreal Protocol: Particularly Remarkable, and Remarkably Particular' (2000–2002) 19 *UCLA Journal of Environmental Law and Policy* 49.

It is arguable that a consequence of focusing on procedural regard in MEA compliance systems is that emphasis is placed on seeking to address states' individual interests, which comes at the cost of strict adherence to the rule of law.[12] For instance, the Montreal compliance system's accountability arrangements for conditional funding accommodate temporary circumvention of binding treaty commitments as funds are paid to countries that are not in substantive compliance but still remain 'in good standing'. These arrangements stand in stark contrast to the traditional rules of state responsibility for breach of an international obligation under general international law.[13] Koskenniemi, for one, critiques the trend toward 'fragmentation' of international law into self-contained regimes, which reflect technocratic 'managerialism', rather than the traditional legal norms of international law.[14]

However, as Hovell observes, due-process models in international law can be designed to promote different values.[15] Hovell distinguishes between three core values underlying the dialogue enabled by due process in international settings: (1) enhancing the accuracy of decision-making outcomes, (2) providing interest representation for those affected by decisions and (3) promoting public accountability in decision making.[16] As noted in Chapter 1, the proceduralization framework in this book reflects due-process values related to interest representation for affected states.[17] These values are primarily concerned with decision makers taking into account diverse needs and perspectives, rather than strict, instrumentalist adherence to a formal set of laws.[18] If states make the substantive choice that legal accuracy is the preferred value for MEA compliance procedures, different procedural design choices will need to be prioritized than if the focus is on enhancing states' opportunities to be heard and understood, and weighing various interest considerations.

Such a shift in the design of compliance frameworks is not necessarily advisable, as insisting on rigid adherence to legal rules in MEA compliance systems may well be self-defeating. As previously mentioned, non-compliance is often attributable to Parties' technical, financial and regulatory capacity constraints rather than deliberate contravention,[19] and taking an inflexible stance on non-compliance will

12 See, eg, J Klabbers, 'Compliance Procedures' in D Bodansky, J Brunnée and E Hey (eds), *The Oxford Handbook of International Environmental Law* (Oxford University Press, 2007) 995, 1007–1009.

13 M Koskenniemi, 'Breach of Treaty or Non-Compliance? Reflections on the Enforcement of the Montreal Protocol' (1992) 3(1) *Yearbook of International Environmental Law* 123, 134.

14 See, eg, M Koskenniemi, 'The Politics of International Law – 20 Years Later' (2009) 20(1) *European Journal of International Law* 7, 15; M Koskenniemi, 'The Fate of Public International Law: Between Technique and Politics' (2007) 70(1) *The Modern Law Review* 1.

15 D Hovell, *The Power of Process: The Value of Due Process in Security Council Sanctions Decision-Making* (Oxford University Press, 2016).

16 *Ibid* 63ff.

17 See Chapter 1, section 2.

18 Hovell, *The Power of Process*, above n 15, 74–76.

19 See, eg, A Chayes and A Handler Chayes, *The New Sovereignty: Compliance with International Regulatory Agreements* (Harvard University Press, 1998) 14.

do little to overcome or address these concerns. Furthermore, in a consent-based context, inflexible approaches may undermine goodwill toward co-operation and increase the risk of treaty exit, as has at times occurred in MEAs.[20] Thus, despite possible objections from a rule of law perspective,[21] for both normative and practical reasons, procedural regard is a desirable focus for MEA compliance systems.

3 Lessons learned for future procedural design in MEA compliance systems

This section reflects upon the insights into procedural design gained from the foregoing synthesis and analysis, and applies these lessons to the emerging MEA compliance apparatus under the Paris Agreement – the successor to the Kyoto Protocol. The Paris Agreement entered into force in November 2016,[22] but does not commence until 2020, and the future development of its modalities, procedures and guidelines may be valuably informed by insights from the proceduralization framework. This discussion reinforces the practical significance of this framework, in addition to its utility as an evaluative tool.

The Paris compliance framework, like other recently negotiated MEA compliance systems,[23] privileges a facilitative, rather than an enforcement-based, approach to non-compliance. I argue that adopting a comprehensive suite of procedures for accountability, transparency, participation, reason giving and review is important to ensure that such facilitative compliance processes are procedurally constrained, and to minimize their vulnerability to being strongly influenced by power imbalances between the Parties. Developing extensive due-process guarantees and minimizing pockets of weakly constrained decision making will help to transform the procedural

20 For example, in 1992 Iceland denounced the International Convention for the Regulation of Whaling, and did not rejoin until 2002: D D Caron, 'The International Whaling Commission and the North Atlantic Marine Mammal Commission: The Institutional Risks of Coercion in Consensual Structures' (1995) 89 *American Journal of International Law* 154, 160–161. See also the discussion of the UAE's exit from CITES in Chapter 6, section 3.

21 On the relationship between the rule of law and administrative procedures in the Montreal compliance system, see further A Huggins, 'Administrative Procedures and Rule of Law Values in the Montreal Compliance System' in C Voigt (ed), *The Environment in International Courts and Tribunals: Questions of Legitimacy* (Cambridge University Press, forthcoming).

22 *Paris Agreement to the United Nations Framework Convention on Climate Change*, opened for signature 12 December 2015 (entered into force 4 November 2016) (in UNFCCC, Report of the Conference of the Parties on its Twenty-First Session, Addendum, UN Doc FCCC/CP/2015/10/Add.1, 29 Jan 2016) ('Paris Agreement').

23 See, eg, the Minamata compliance system under the *Minamata Convention on Mercury*, opened for signature 10 October 2013, entry into force 16 August 2017, art 15; J Templeton and P Kohler, 'Implementation and Compliance under the Minamata Convention on Mercury' (2014) 23(2) *Review of European Community and International Environmental Law* 211.

opportunities evident in this nascent phase of the Paris Agreement into a robust procedural framework.

The Paris Agreement is a global, legally binding instrument with the same core obligations for all countries. Its architecture is durable, creating a framework for progressively stronger action over time.[24] Before proceeding to analyze the Agreement's compliance apparatus, it is pertinent to briefly canvass key differences between the Kyoto Protocol and the Paris Agreement with respect to mitigation commitments, which are the typical focus of non-compliance proceedings.

Significantly, the Kyoto Protocol's differentiated mitigation obligations for developed and developing countries, which mean that developing states do not have binding emissions reductions targets, have not been replicated under the Paris Agreement. The latter agreement privileges universal participation and bottom-up 'self-differentiation', through which states select their own mitigation targets.[25] Under Article 2 of the Paris Agreement, there is a collective general obligation on *all* state Parties to hold 'the increase in the global average temperature to well below 2°C above pre-industrial levels and to pursue efforts to limit the temperature increase to 1.5°C above pre-industrial levels'.[26] This new self-differentiated approach means that states are required to specify their own nationally determined contributions (NDC) to the global climate-change response, which will be progressively strengthened,[27] rather than having binding mitigation targets specified in the Paris Agreement itself.[28] The Agreement does, however, include binding obligations for all states to prepare, communicate, maintain and periodically report national contributions, and pursue domestic mitigation measures,[29] and for developed states to continue to provide financial resources to assist developing countries.[30] Hence, the binding obligations for achieving the Agreement's long-term temperature goal in Article 2 are procedurally oriented, rather than substantive, in nature.

The Paris Agreement provides for four key oversight mechanisms for state reports submitted through the formal reporting system – the transparency framework for action and support,[31] multilateral consideration of progress,[32] global stocktakes[33]

24 D Bodansky, 'The Paris Climate Change Agreement: A New Hope?' (2016) 110(2) *American Journal of International Law* 289, 290–291.

25 L Rajamani, 'The Devilish Details: Key Legal Issues in the 2015 Climate Negotiations' (2015) 78(5) *Modern Law Review* 826, 852.

26 Paris Agreement, above n 22, art 2.

27 *Ibid* art 4(3).

28 L Rajamani, 'Ambition and Differentiation in the 2015 Paris Agreement: Interpretative Possibilities and Underlying Politics' (2016) 65(2) *International and Comparative Law Quarterly* 493, 497.

29 Paris Agreement, above n 22, art 4(2) and 4(3). Under art 4(3), Parties are required to communicate their contributions every five years.

30 *Ibid* art 9(1).

31 *Ibid* art 13.

32 *Ibid* art 13(11).

33 *Ibid* art 14.

and the formal non-compliance process.[34] All four will play an important role in promoting states' accountability for implementing their commitments under the Paris Agreement, and thus warrant consideration.

The transparency framework for action and support recognizes the need for greater transparency in relation to state-provided information on mitigation progress and support in order to build mutual trust and confidence between the Parties, and to promote effective implementation.[35] In addition to a requirement for all states to report on their progress toward implementing and achieving their NDCs,[36] developed states are required to report on the financial, technology transfer and capacity-building support they provide to developing countries.[37] Technical expert review will consider the support provided by a Party (where relevant), and its implementation and achievement of its NDC, paying particular attention to the respective national capabilities and circumstances of developing-country Parties.[38] The transparency framework will build upon the reporting and review processes developed under the UNFCCC,[39] which since the 2010 Cancun Agreements include processes known as international assessment and review (IAR) and international consultation and analysis (ICA) for developed and developing countries, respectively. IAR is a 'robust, rigorous and transparent process' to promote 'comparability and confidence' that applies to developed countries' greenhouse gas inventories, biennial reports and national communications. ICA is a type of 'non-intrusive, non-punitive' review that is 'respectful of national sovereignty' while increasing 'transparency of mitigation actions and their effects' that applies to developing countries' biennial update reports.[40] The Paris Agreement specifies that the modalities, procedures and guidelines for transparency to be developed by the CMP should be 'common',[41] thus tempering the degree of differentiation between reporting and review processes for developed and developing states that is currently evident under the IAR and ICA arrangements.

The information reported by states will be relevant to a 'facilitative, multilateral consideration of progress' with respect to states' implementation and achievement

34 *Ibid* art 15.
35 *Ibid* art 13(1) and 13(7).
36 *Ibid* art 13(7).
37 *Ibid* art 13(9). Developing countries should also report on the financial, technology transfer and capacity-building support needed and received: *ibid* art 13(10). Art 9(1) obligates developed countries to provide financial resources to developing countries, and art 9(3) urges them to take the lead in mobilizing and progressively increasing funds for climate finance.
38 *Ibid* art 13(11) and (12). The transparency framework will be implemented flexibly for those developing countries that require this in the light of their capacities: *ibid.*
39 *Ibid* art 13(3).
40 Decision 1/CP.16, Report of the Conference of the Parties on its Sixteenth Session, Held in Cancun from 29 November to 10 December 2010, FCCC/CP/2010/7/Add.1, 15 Mar 2011, paras 44, 63; Decision 2/CP.17, Outcome of the Work of the Ad Hoc Working Group on Long-Term Cooperative Action under the Convention, FCCC/CP/2011/9/Add.1, 15 Mar 2012, paras 23–31 and Annex II (for IAR); and paras 56–62 (for ICA).
41 Paris Agreement, above n 22, art 13(13).

of their NDCs and, for developed-country Parties, their efforts with respect to providing financial support for developing countries.[42] The specifics of this mechanism are yet to be developed,[43] but again appear likely to draw upon existing oversight processes for developing and developed states' mandatory reports under the UNFCCC, which do not include strong consequences for non-compliance. The Cancun Agreements designated the Subsidiary Body for Implementation (SBI) as the political body responsible for overseeing the reporting and review processes for both developed and developing countries under IAR and ICA. The SBI undertakes 'multilateral assessment' for the former, and a 'facilitative sharing of views' for the latter.[44] The outcomes of the multilateral assessment phase of IAR are a set of SBI conclusions, which are forwarded to 'relevant bodies under the Convention as appropriate'.[45] A 'summary report and a record of the facilitative sharing of views' are the only outcomes required for the multilateral phase of ICA.[46] The experience of the SBI in supervising the IAR and ICA processes is likely to inform the development of the Paris Agreement's 'multilateral, facilitative consideration of progress'.[47] Significantly, the SBI's supervisory processes for both IAR and ICA rely on public 'naming and shaming', rather than concrete consequences for Parties in non-compliance.[48]

The global stocktake process will also have a facilitative nature, and will focus on collective progress, rather than individual state accountability. The stocktake will be held every five years, beginning in 2023, and its aim is to assess 'collective progress towards achieving the purpose of this Agreement and its long term goals', which suggests that it will contribute to the Agreement's ratcheting mechanism.[49] The stocktake is to be conducted in a facilitative manner 'in the light of equity and the best available science', and its focus will be on mitigation, adaptation and the means of implementation and support.[50] Given the absence of quantifiable, time-bound targets in the Paris Agreement for mitigation, finance, technology support and capacity-building, and the emphasis on collective progress, there appears to be limited scope for the global stocktake process to hold individual states to account for failure to meet national targets.[51]

42 *Ibid* art 13(11).
43 It is as yet unclear how the 'facilitative, multilateral consideration of progress' will be conducted, what its outcomes will be and how it will feed into the global stocktake: Rajamani, 'Ambition and Differentiation in the 2015 Paris Agreement', above n 28, 503.
44 Decision 2/CP.17, above n 40, Annexes II and IV.
45 *Ibid* Annex II, paras 11 and 12.
46 *Ibid* Annex IV, para 8.
47 A Zahar, 'A Bottom-Up Compliance Mechanism for the Paris Agreement', *Chinese Journal of Environmental Law* (forthcoming, 2017) 19 <https://papers.ssrn.com/sol3/papers.cfm?abstract_id=2912560>.
48 S Oberthür, 'Options for a Compliance Mechanism in a 2015 Climate Agreement' (2014) 4 *Climate Law* 30, 43.
49 Paris Agreement, above n 22, art 14(1) and 14(3).
50 *Ibid* art 14(1).
51 Rajamani, 'Ambition and Differentiation in the 2015 Paris Agreement', above n 28, 504.

Finally, Article 15 provides for a mechanism to facilitate implementation of, and compliance with, the Paris Agreement. This Article specifies that the compliance committee will be expert-based, facilitative, non-adversarial and non-punitive, and is required to 'pay particular attention to the respective national capabilities and circumstances of Parties'.[52] Accordingly, the compliance mechanism departs from the Kyoto Protocol's Enforcement Branch model, which included economic sanctions as a potential non-compliance response measure, and is anticipated to more closely approximate the design of the Facilitative Branch.[53] This mechanism is likely to rely on peer and public pressure as a means of influencing state behavior,[54] without recourse to enforcement mechanisms.

The proceduralization framework provides a number of valuable insights for the future development of the Paris Agreement's compliance apparatus. Perhaps most significantly, all of the oversight mechanisms incorporated in the Paris Agreement are hampered by a lack of robust consequences for deficient performance, which threatens to undermine the efficacy of these accountability processes.[55] The absence of quantifiable, time-bound mitigation and finance commitments in the Agreement limits possibilities for holding individual states to account for poor performance, impeding accountability aims. Furthermore, the exclusive emphasis on a facilitative, non-punitive approach to overseeing states' performance is significant, as an important aspect of accountability is the potential imposition of 'sanctions or other remedies for deficient performance'.[56] A paucity of such consequences is likely to create a lacuna in the procedural framework that is vulnerable to exploitation by states, as has been evident in other pockets of global environmental governance with inadequate accountability mechanisms.[57] If the Paris Agreement proceeds without incorporating any consequences or intrusive measures for non-compliance with legal obligations, the role of IEL in shaping state behavior may need to be reconsidered,[58] which is concerning given the high stakes of failure to take effective action on climate change.

52 Paris Agreement, above n 22, art 15(2).

53 Cf Voigt, who argues that the Paris Agreement's compliance committee should consist of two branches – a compliance branch and an implementation branch: C Voigt, 'The Compliance and Implementation Mechanism of the Paris Agreement' (2016) 25(2) *Review of European Community and International Environmental Law* 161, 166–169.

54 Bodansky, 'The Paris Climate Change Agreement', above n 24, 311.

55 See further A Huggins and M S Karim, 'Differential Treatment and Substantive and Procedural Regard in the International Climate Change Regime' (2016) 5(2) *Transnational Environmental Law* 427, 441–443.

56 Stewart, 'Remedying Disregard in Global Regulatory Governance', above n 10, 253.

57 In this regard, a cautionary tale is provided by the International Maritime Organization's (IMO's) regime for reducing greenhouse gas emissions from marine bunker fuels, which has proved ineffective in promoting developing countries' interests, largely because it does not incorporate adequate accountability mechanisms: Huggins and Karim, 'Differential Treatment and Substantive and Procedural Regard', above n 55, 442–448.

58 See, eg, H van Asselt, 'International Climate Change Law in a Bottom-up World' (2016) 26 *Questions of International Law* 5, 10.

Second, as discussed in Chapter 5, in the Kyoto compliance system, ERTs have assumed extensive ad hoc and informal facilitative roles for which they are arguably insufficiently equipped. Given the facilitative processes associated with multilateral consideration of process, global stocktakes and non-compliance under the Paris Agreement, it is appropriate for technical expert reviewers to assume narrower technical roles, in accordance with their expertise, rather than serving as an additional tier of facilitative review.[59] This suggested role clarification underscores the desirability of incorporating further procedural safeguards into the expert review process. Such safeguards should include guidelines for enhancing training, political independence, the consistency and transparency of decision-making processes, reason-giving requirements and ensuring that the rules and guidelines for expert review do not include any references to facilitative roles. This would clarify that the review role is primarily technical, and would provide scope for the Compliance Committee to assume a more comprehensive facilitative role, safeguarded by due-process guarantees.[60]

Third, the relationship between the various oversight mechanisms under the Paris Agreement needs to be clarified. At present, it is unclear how technical expert review processes will inform the multilateral consideration of progress, the global stocktake and the non-compliance mechanism.[61] In clarifying this relationship, it is important to bear in mind the tendency for decision-making power to shift to zones where procedural constraints are thinner, and political negotiations prevail, which was evident in the preceding case study analysis. In order to limit the potential political influences on compliance decision making, it is preferable for procedurally constrained expert review processes to feed directly into the non-compliance process, rather than having the facilitative processes associated with multilateral consideration of progress and/or the global stocktake as additional intermediary steps.

Finally, due-process guarantees for the Compliance Committee need to be further developed. Such safeguards can be modeled on the Facilitative Branch approach, which includes requirements for adequate notice of alleged non-compliance, an opportunity to respond orally and in writing, a right to representation, a right to an oral hearing, reason-giving requirements and measures to promote the impartiality of expert decision makers.[62] This type of proceduralized compliance

59 Cf Zahar, who argues for a delay in developing a compliance mechanism under Article 15 due to, inter alia, the extensive facilitative roles already played by ERTs: Zahar, 'A Bottom-Up Compliance Mechanism for the Paris Agreement', above n 47.

60 See also A Huggins, 'The Desirability of Depoliticization: Compliance in the International Climate Regime' (2015) 4(1) *Transnational Environmental Law* 101, 120–123.

61 M Doelle, 'Reviewing Implementation and Compliance under the Paris Climate Agreement: Preliminary Thoughts on Process Design for Articles 13-15' (2016) 3–4 <https://papers.ssrn.com/sol3/papers.cfm?abstract_id=2762821>. Articles 13(5) and 13(6) state that the transparency framework is intended to inform, inter alia, the global stocktake, yet it remains unclear how this will occur.

62 See further Chapter 5.

framework holds significant promise for enhancing regard for affected states' interests, and minimizing the risk that compliance decisions are shaped by power politics. As the experience of the Montreal compliance system demonstrates, such opportunities for procedural regard can be buttressed by the availability of financial and technical assistance to support non-compliant developing states' return to compliance,[63] which promotes both procedural and substantive regard for affected states' interests. The prospect of more intrusive consequences for non-compliance if facilitative measures are not successful is beneficial to reinforce accountability. Thus, the proceduralization framework yields valuable insights for the future design and development of the Paris Agreement, compliance with which is vital for reducing the risk of severe and irreversible climate change.

4 Conclusion

Procedural design that places constraints upon power politics and creates opportunities for affected interests to be voiced and considered is highly desirable in MEA compliance systems. Despite the challenges associated with practical implementation, administrative procedures have a demonstrated, if as yet under-utilized, potential to discipline politics and enhance procedural regard in these settings. Adopting administrative procedures is important to signal respect for states' autonomy, and diverse capacities and circumstances, and to tailor and justify appropriate compliance solutions that are responsive to the reasons for states' non-compliance. In a global environmental governance context shaped by North–South power imbalances, and predicated on states' consent, participation and goodwill, leveling the playing field for less powerful states in compliance processes is important for the political and environmental success of MEAs. Given the indispensable role that compliance systems play in promoting the achievement of the critical environmental goals of IEL, the optimal design of MEA compliance systems deserves to be prioritized.

63 See also Huggins and Karim, 'Differential Treatment and Substantive and Procedural Regard', above n 55, 444–448; C F Sabel and D G Victor, 'Governing Global Problems under Uncertainty: Making Bottom-up Climate Policy Work' (2015) *Climatic Change* 1, 10.

Bibliography

A *Articles/Books/Reports*

Abbott, K W, P Genschel, D Snidal and B Zangl, 'Orchestration: Global Governance through Intermediaries' in K W Abbott, P Genschel, D Snidal and B Zangl (eds), *International Organizations as Orchestrators* (Cambridge University Press, 2015) 3

Abbott, K W, R O Keohane, A Moravcsik, A-M Slaughter and D Snidal, 'The Concept of Legalization' (2000) 54(3) *International Organization* 401

Abbott, K W, D Levi-Faur and D Snidal, 'Theorizing Regulatory Intermediaries: The RIT Model' (2017) 670 *Annals of the American Academy of Political and Social Science* 14

Abbott, K W, and D Snidal, 'Hard and Soft Law in International Governance' (2000) 54(3) *International Organization* 421

Alam, S, S Atapattu, C G Gonzalez and J Razzaque (eds), *International Environmental Law and the Global South* (Cambridge University Press, 2015)

Alter, K J, *The New Terrain of International Law: Courts, Politics, Rights* (Princeton University Press, 2014)

Anghie, A, *Imperialism, Sovereignty and the Making of International Law* (Cambridge University Press, 2005)

Bafundo, N E, 'Compliance with the Ozone Treaty: Weak States and the Principle of Common but Differentiated Responsibility' (2005–2006) 21 *American University International Law Review* 461

Bankobeza, G M, *Ozone Protection: The International Legal Regime* (Eleven International Publishing, 2005)

Bankobeza, S, E Maruma Mrema and K Ramakrishna, 'International Environmental Diplomacy and Negotiations' in T Kuokkanen, E Couzens, T Honkonen and M Lewis (eds), *International Environmental Law-making and Diplomacy: Insights and Overviews* (Routledge, 2016) 83

Bauer, S, P-O Busch and B Siebenhüner, 'Treaty Secretariats in Global Environmental Governance' in F Bierman, B Siebenhüner and A Schreyögg (eds), *International Organizations in Global Environmental Governance* (Routledge, 2009) 174

Bayles, M D, *Procedural Justice: Allocating to Individuals* (Kluwer International, 1990)

Benvenisti, E, *The Law of Global Governance* (Hague Academy of International Law, 2014)

Berntsen, T, J Fuglestvedt and F Stordal, 'Reporting and Verification of Emissions and Removals of Greenhouse Gases' in O S Stokke, J Hovi and G Ulfstein (eds), *Implementing the Climate Regime: International Compliance* (Earthscan, 2005) 85

Beyerlin, U, and T Marauhn, *International Environmental Law* (Hart Publishing, 2011)

Bianchi, A, 'On Power and Illusion: The Concept of Transparency in International Law' in A Bianchi and A Peters (eds), *Transparency in International Law* (Cambridge University Press, 2013) 1

Bignami, F, 'From Expert Administration to Accountability Network: A New Paradigm for Comparative Administrative Law' (2011) 59 *American Journal of Comparative Law* 859

Black, J, 'Constructing and Contesting Legitimacy and Accountability in Polycentric Regulatory Regimes' (2008) 2 *Regulation and Governance* 136

Bodansky, D, 'Is There an International Environmental Constitution?' (2009) 16(2) *Indiana Journal of Global Legal Studies* 565

Bodansky, D, 'Legitimacy' in D Bodansky, J Brunnée and E Hey (eds), *The Oxford Handbook of International Environmental Law* (Oxford University Press, 2007) 704

Bodansky, D, 'Legitimacy in International Law and International Relations' in J L Dunoff and M A Pollack (eds), *Interdisciplinary Perspectives on International Law and International Relations: The State of the Art* (Cambridge University Press, 2012) 321

Bodansky, D, *The Art and Craft of International Environmental Law* (Harvard University Press, 2009)

Bodansky, D, 'The Paris Climate Change Agreement: A New Hope?' (2016) 110(2) *American Journal of International Law* 289

Borras, S, 'Comparative Analysis of Selected Compliance Procedures under Multilateral Environmental Agreements' in S Maljean-Dubois and L Rajamani (eds), *Implementation of International Environmental Law* (Martinus Nijhoff Publishers, 2011) 319

Bovens, M, 'Analysing and Assessing Accountability: A Conceptual Framework' (2007) 13 *European Law Journal* 447

Brack, D, 'Monitoring the Montreal Protocol' (2003) *Verification Yearbook* 209

Brown Weiss, E, 'The Five International Treaties: A Living History' in E Brown Weiss and H K Jacobson (eds), *Engaging Countries: Strengthening Compliance with International Environmental Accords* (MIT Press, 1998) 89

Brunnée, J, 'Climate Change and Compliance and Enforcement Processes' in R Rayfuse and S V Scott (eds), *International Law in the Era of Climate Change* (Edward Elgar, 2012) 290

Brunnée, J, 'Common Areas, Common Heritage, and Common Concern' in D Bodansky, J Brunnée and E Hey (eds), *The Oxford Handbook of International Environmental Law* (Oxford University Press, 2007) 550

Brunnée, J, 'Compliance Control' in G Ulfstein, T Marauhn and A Zimmermann (eds), *Making Treaties Work: Human Rights, Environment and Arms Control* (Cambridge University Press, 2007) 373

Brunnée, J, 'COPing with Consent: Law-Making under Multilateral Environment Agreements' (2002) 15(1) *Leiden Journal of International Law* 1

Brunnée, J, 'Enforcement Mechanisms in International Law and International Environmental Law' in U Beyerlin, P-T Stoll and W Rudiger (eds), *Ensuring Compliance with Multilateral Environmental Agreements: A Dialogue Between Practitioners and Academia* (Brill Martinus Nijhoff Publishers, 2006) 1

Brunnée, J, 'International Environmental Law and Community Interests: Procedural Aspects' in E Benvenisti and G Nolte (eds), *Community Obligations in International*

Law (forthcoming) <https://papers.ssrn.com/sol3/papers.cfm?abstract_id=2784701>

Brunnée, J, 'International Legal Accountability through the Lens of the Law of State Responsibility' (2005) 36 *Netherlands Yearbook of International Law* 21

Brunnée, J, 'Multilateral Environmental Agreements and the Compliance Continuum' in G Winter (ed), *Multilevel Governance of Global Environmental Change: Perspectives from Science, Sociology and the Law* (Cambridge University Press, 2006) 387

Brunnée, J, and E Hey, 'Transparency and International Environmental Institutions' in A Bianchi and A Peters (eds), *Transparency in International Law* (Cambridge University Press, 2013) 23

Brunnée, J, and S J Toope, *Legitimacy and Legality in International Law: An Interactional Account* (Cambridge University Press, 2010)

Buchanan, A, and R O Keohane, 'The Legitimacy of Global Governance Institutions' (2006) 20(4) *Ethics & International Affairs* 405

Bulmer, J, 'Compliance Regimes in Multilateral Environmental Agreements' in J Brunnée, M Doelle and L Rajamani (eds), *Promoting Compliance in an Evolving Climate Regime* (Cambridge University Press, 2012) 55

Cardesa-Salzmann, A, 'Constitutionalising Secondary Rules in Global Environmental Regimes: Non-Compliance Procedures and the Enforcement of Multilateral Environmental Agreements' (2011) 24(1) *Journal of Environmental Law* 103

Caron, D D, 'The International Whaling Commission and the North Atlantic Marine Mammal Commission: The Institutional Risks of Coercion in Consensual Structures' (1995) 89 *American Journal of International Law* 154

Cassese, S, 'A Global Due Process of Law?' in G Anthony, J-B Auby, J Morison and T Zwart (eds), *Values in Global Administrative Law* (Hart Publishing, 2011) 17

Cassese, S (ed), *Research Handbook on Global Administrative Law* (Edward Elgar, 2016)

Charnovitz, S, 'Nongovernmental Organizations and International Law' (2006) 100 *American Journal of International Law* 348

Chayes, A, and A Handler Chayes, *The New Sovereignty: Compliance with International Regulatory Agreements* (Harvard University Press, 1998)

Chesterman, S, 'Globalisation and Public Law: A Global Administrative Law?' in J Farrall and K Rubenstein (eds), *Sanctions, Accountability and Governance in a Globalised World* (Cambridge University Press, 2009) 75

Chimni, B S, 'Co-option and Resistance: Two Faces of Global Administrative Law' (2005) 37 *New York University Journal of International Law and Politics* 799

Chimni, B S, 'Global Administrative Law: Winners and Losers' (Paper presented at the NYU Law School Global Administrative Law Conference, 22–23 April 2005) <http://iilj.org/global_adlaw/documents/ChimniPaper.pdf>

Churchill, R R, and G Ulfstein, 'Autonomous Institutional Arrangements in Multilateral Environmental Agreements: A Little-Noticed Phenomenon in International Law' (2000) 94 *American Journal of International Law* 623

Cooney, R, and M Abensperg-Traun, 'Raising Local Community Voices: CITES, Livelihoods and Sustainable Use' (2013) 22(3) *Review of European Community and International Environmental Law* 301

Couzens, E, *Whales and Elephants in International Conservation Law and Politics: A Comparative Study* (Routledge, 2014)

Crawford, J, 'Marbury v. Madison at the International Level' (2004) 36 *George Washington International Law Review* 505

Davies, P G, 'Non-Compliance – A Pivotal or Secondary Function of CoP Governance?' (2013) 15 *International Community Law Review* 77 de Burca, G, R O Keohane and C F Sabel, 'Global Experimentalist Governance' (2014) 44(3) *British Journal of Political Science* 477

de Wet, E, 'Judicial Review as an Emerging Principle of Law and its Implications for the International Court of Justice' (2000) 47(2) *Netherlands International Law Review* 181

DeSombre, E, 'The Experience of the Montreal Protocol: Particularly Remarkable, and Remarkably Particular' (2000–2002) 19 *UCLA Journal of Environmental Law and Policy* 49

Doelle, M, 'Compliance and Enforcement in the Climate Change Regime', in E J Hollo, K Kulovesi and M Mehling (eds), *Climate Change and the Law* (Springer, 2012) 165

Doelle, M, 'Early Experience with the Kyoto Compliance System: Possible Lessons for MEA Compliance System Design' (2010) 1(2) *Climate Law* 237

Doelle, M, 'Experience with the Facilitative and Enforcement Branches of the Kyoto Compliance System' in J Brunnée, M Doelle and L Rajamani (eds), *Promoting Compliance in an Evolving Climate Regime* (Cambridge University Press, 2012) 102

Doelle, M, 'Reviewing Implementation and Compliance under the Paris Climate Agreement: Preliminary Thoughts on Process Design for Articles 13-15' (2016) <https://papers.ssrn.com/sol3/papers.cfm?abstract_id=2762821>

Donaldson, M, and B Kingsbury, 'The Adoption of Transparency Policies in Global Governance Institutions: Justifications, Effects, and Implications' (2013) 9 *Annual Review of Law and Social Science* 119

Downs, G W, D M Rocke and P N Barsoom, 'Is the Good News about Compliance Good News about Cooperation?' (1996) 50(3) *International Organization* 379

Dubash, N, and B Morgan, 'The Embedded Regulatory State: Between Rules and Deals' in N Dubash and B Morgan (eds), *The Rise of the Regulatory State of the South: Infrastructure and Development in Emerging Economies* (Oxford University Press, 2013) 279

Dubash, N, and B Morgan (eds), *The Rise of the Regulatory State of the South: Infrastructure and Development in Emerging Economies* (Oxford University Press, 2013)

Duffy, R, 'Global Environmental Governance and North–South Dynamics: The Case of CITES' (2013) 31 *Environment and Planning C: Government and Policy* 222

Ebbesson, J, 'Public Participation' in D Bodansky, J Brunnée and E Hey (eds), *The Oxford Handbook of International Environmental Law* (Oxford University Press, 2007) 681

Epiney, A, 'The Role of NGOs in the Process of Ensuring Compliance with MEAs' in U Beyerlin, P-T Stoll and W Rudiger (eds), *Ensuring Compliance with Multilateral Environmental Agreements: A Dialogue Between Practitioners and Academia* (Brill Martinus Nijhoff Publishers, 2006) 319

Esty, D C, 'Environmental Protection in the Information Age' (2004) 79 *New York University Law Review* 115

Esty, D C, 'Good Governance at the Supranational Level: Globalizing Administrative Law' (2006) 115 *Yale Law Journal* 1490

Fisher, E, *Risk Regulation and Administrative Constitutionalism* (Hart Publishing, 2010)

Fisher, E, 'Transparency and Administrative Law: A Critical Evaluation' (2010) 63 *Current Legal Problems* 272

Fitzmaurice, M, 'Non-Compliance Procedures and the Law of Treaties' in T Treves et al (eds), *Non-Compliance Procedures and Mechanisms and the Effectiveness of International Environmental Agreements* (Cambridge University Press, 2009) 453

Fitzmaurice, M A, and C Redgwell, 'Environmental Non-compliance Procedures and International Law' (2000) 31 *Netherlands Yearbook of International Law* 35

Fodella, A, 'Structural and Institutional Aspects of Non-Compliance Mechanisms' in T Treves et al (eds), *Non-Compliance Procedures and Mechanisms and the Effectiveness of International Environmental Agreements* (TMC Asser Press, 2009) 355

Franck, T, *Fairness in International Law and Institutions* (Oxford University Press, 1995)

Fransen, T, 'Enhancing Today's MRV Framework to Meet Tomorrow's Needs: The Role of National Communications and Inventories' (World Resources Institute, 2009) 8 <www.wri.org/sites/default/files/national_communications_mrv.pdf>

Fuchs, C, 'Convention on International Trade in Endangered Species of Wild Fauna and Flora (CITES) – Conservation Efforts Undermine the Legality Principle' (2008) 9(11) *German Law Journal* 1565

Galligan, D J, *Due Process and Fair Procedures: A Study of Administrative Procedures* (Clarendon Press, 1996)

Gareau, B J, *From Precaution to Profit: Contemporary Challenges to Environmental Protection in the Montreal Protocol* (Yale University Press, 2013)

Gathii, J T, 'TWAIL: A Brief History of its Origins, its Decentralized Network and a Tentative Bibliography' (2011) 3(1) *Trade, Law and Development* 26

Gehring, T, 'Treaty-Making and Treaty Evolution' in D Bodansky, J Brunnée and E Hey (eds), *The Oxford Handbook of International Environmental Law* (Oxford University Press, 2007) 467

Gillespie, A, *Climate Change, Ozone Depletion, and Air Pollution: Legal Commentaries with Policy and Scientific Considerations* (Martinus Nijhoff Publishers, 2006)

Gillespie, A, 'Implementation and Compliance Concerns in International Environmental Law: The State of the Art within Three International Regimes' (2003) 7 *New Zealand Journal of Environmental Law* 53

Goeteyn, N, and F Maes, 'Compliance Mechanisms in Multilateral Environmental Agreements: An Effective Way to Improve Compliance?' (2011) 10 *Chinese Journal of International Law* 791

Grant, R W, and R O Keohane, 'Accountability and Abuses of Power in World Politics' (2005) 99(1) *American Political Science Review* 29

Greene, O, 'The System for Implementation Review in the Ozone System' in D G Victor, K Raustiala and E B Skolnikoff (eds), *The Implementation and Effectiveness of International Environmental Commitments* (MIT Press, 1998) 89

Gupta, A, 'Transparency as Contested Political Terrain: Who Knows What about the Global GMO Trade and Why Does It Matter?' (2010) 10(3) *Global Environmental Politics* 32

Gupta, A, and M Mason (eds), *Transparency in Global Environmental Governance: Critical Perspectives* (MIT Press, 2014)

Gupta, J, 'Developing Countries: Trapped in the Web of Sustainable Development Governance' in O Dilling, M Herberg and G Winter (eds), *Transnational Administrative Rule-Making: Performance, Legal Effects, and Legitimacy* (Hart Publishing, 2011) 305

Guzman, A T, 'A Compliance-based Theory of International Law' (2002) 90 *California Law Review* 1823

Guzman, A T, *How International Law Works: A Rational Choice Theory* (Oxford University Press, 2008)

Handl, G, 'Compliance Control Mechanisms and International Environmental Obligations' (1997) 5 *Tulane Journal of International and Comparative Law* 29

Harlow, C, 'Accountability as a Value for Global Governance and Global Administrative Law' in G Anthony, J-B Auby, J Morison and T Zwart (eds), *Values in Global Administrative Law* (Hart Publishing, 2011) 173

Harlow, C, *Accountability in the European Union* (Oxford University Press, 2002)

Harlow, C, 'Global Administrative Law: The Quest for Principles and Values' (2006) 17(1) *The European Journal of International Law* 187

Hawkins, D G, D A Lake, D L Nielson and M J Tierney, 'Delegation under Anarchy: States, International Organizations and Principal-Agent Theory' in D G Hawkins, D A Lake, D L Nielson and M J Tierney (eds), *Delegation and Agency in International Organizations* (Cambridge University Press, 2006) 3

Herold, A, 'Experiences with Articles 5, 7, and 8: Defining the Monitoring, Reporting and Verification System under the Kyoto Protocol' in J Brunnée, M Doelle and L Rajamani (eds), *Promoting Compliance in an Evolving Climate Regime* (Cambridge University Press, 2012) 122

Hey, E, 'International Institutions' in D Bodansky, J Brunnée and E Hey (eds), *The Oxford Handbook of International Environmental Law* (Oxford University Press, 2007) 749

Hey, E, 'The Climate Change Regime: An Enviro-Economic Problem and International Administrative Law in the Making' (2001) 1 *International Environmental Agreements: Politics, Law and Economics* 75

Honkonen, T, 'The Development of the Principle of Common but Differentiated Responsibilities and its Place in International Environmental Regimes' in T Kuokkanen, E Couzens, T Honkonen and M Lewis (eds), *International Environmental Law-making and Diplomacy: Insights and Overviews* (Routledge, 2016) 160

Hovell, D, 'Due Process in the United Nations' (2016) 110(1) *American Journal of International Law* 1

Hovell, D, *The Power of Process: The Value of Due Process in Security Council Sanctions Decision-Making* (Oxford University Press, 2016)

Howarth, W, 'Aspirations and Realities under the Water Framework Directive: Proceduralisation, Participation and Practicalities' (2009) 21(3) *Journal of Environmental Law* 391

Huggins, A, 'Administrative Procedures and Rule of Law Values in the Montreal Compliance System' in C Voigt (ed), *The Environment in International Courts and Tribunals: Questions of Legitimacy* (Cambridge University Press, forthcoming)

Huggins, A, *The Desirability of Administrative Proceduralisation: Compliance Rules and Decisions in Multilateral Environmental Agreements* (PhD thesis, UNSW, October 2015)

Huggins, A, 'The Desirability of Depoliticization: Compliance in the International Climate Regime' (2015) 4(1) *Transnational Environmental Law* 101

Huggins, A, and M S Karim, 'Differential Treatment and Substantive and Procedural Regard in the International Climate Change Regime' (2016) 5(2) *Transnational Environmental Law* 427

Hunter, D B, 'The Emerging Norm of Transparency in International Environmental Governance' in P Ala'i and R G Vaughn (eds), *Research Handbook on Transparency* (Edward Elgar, 2014) 343

Hunter, D, J Salzman and D Zaelke, *International Environmental Law and Policy* (Foundation Press, 4th ed, 2011)

Hutton, J M, and B Dickson, *Endangered Species, Threatened Convention: The Past, Present and Future of CITES, the Convention on International Trade in Endangered Species of Wild Fauna and Flora* (Earthscan, 2000)

Kingsbury, B, 'Global Environmental Governance as Administration: Implications for International Law' in D Bodansky, J Brunnée and E Hey (eds), *The Oxford Handbook of International Environmental Law* (Oxford University Press, 2007) 63

Kingsbury, B, 'The Concept of Compliance as a Function of Competing Conceptions of International Law' (1998) 19 *Michigan Journal of International Law* 345

Kingsbury, B, 'The Concept of "Law" in Global Administrative Law' (2009) 20(1) *European Journal of International Law* 23

Kingsbury, B, and L Casini, 'Global Administrative Law Dimensions of International Organizations Law' (2009) 6 *International Organizations Law Review* 319

Kingsbury, B, N Krisch and R B Stewart, 'The Emergence of Global Administrative Law' (2005) 68 *Law and Contemporary Problems* 15

Kishore, P, 'A Comparative Analysis of Secretariats Created under Select Treaty Regimes' (2011) 45 *The International Lawyer* 1051

Klabbers, J, *An Introduction to International Institutional Law* (Cambridge University Press, 2nd ed, 2009)

Klabbers, J, 'Compliance Procedures' in D Bodansky, J Brunnée and E Hey (eds), *The Oxford Handbook of International Environmental Law* (Oxford University Press, 2007) 995

Klabbers, J, A Peters and G Ulfstein (eds), *The Constitutionalization of International Law* (Oxford University Press, 2009)

Koh, H H, 'Transnational Legal Process' (1996) 75 *Nebraska Law Review* 181

Koppell, J, 'Pathologies of Accountability: ICANN and the Challenge of "Multiple Accountability Disorder"' (2005) 65(1) *Public Administration Review* 94

Koskenniemi, M, 'Breach of Treaty or Non-Compliance? Reflections on the Enforcement of the Montreal Protocol' (1992) 3(1) *Yearbook of International Environmental Law* 123

Koskenniemi, M, 'The Fate of Public International Law: Between Technique and Politics' (2007) 70(1) *The Modern Law Review* 1

Koskenniemi, M, 'The Politics of International Law – 20 Years Later' (2009) 20(1) *European Journal of International Law* 7

Kotze, L J, 'Arguing Global Environmental Constitutionalism' (2012) 1(1) *Transnational Environmental Law* 199

Krisch, N, 'Global Administrative Law and the Constitutional Ambition' in P Dobner and M Loughlin (eds), *The Twilight of Constitutionalism* (Oxford University Press, 2010) 245

Krisch, N, 'The Decay of Consent: International Law in an Age of Global Public Goods' (2014) 108(1) *American Journal of International Law* 1

Krisch, N, and B Kingsbury, 'Introduction: Global Governance and Global Administrative Law in the International Legal Order' (2006) 17(1) *The European Journal of International Law* 1

Kumm, M, 'The Legitimacy of International Law: A Constitutionalist Framework Analysis' (2004) 15 *European Journal of International Law* 907

Lefeber, R, and S Oberthur, 'Key Features of the Kyoto Protocol's Compliance System' in J Brunnée, M Doelle and L Rajamani (eds), *Promoting Compliance in an Evolving Climate Regime* (Cambridge University Press, 2012) 77

Mackenzie, R, 'The Role of Dispute Settlement in the Climate Regime' in J Brunnée, M Doelle and L Rajamani (eds), *Promoting Compliance in an Evolving Climate Regime* (Cambridge University Press, 2012) 395

Madhava Sarma, K, and K N Taddonio, 'The Role of Financial Assistance by the Multilateral Fund in Technology Change to Protect the Ozone Layer' in Z Cerefos, G Contopoulos and G Skalkeas (eds), *Twenty Years of Ozone Decline: Proceedings of the Symposium for the 20th Anniversary of the Montreal Protocol* (Springer, 2009) 441

Marauhn, T, 'Towards a Procedural Law of Compliance Control in International Environmental Relations' (1996) 56(3) *Zeitschrift für Ausländisches Recht und Völkerrecht* 696

Marks, S, 'Naming Global Administrative Law' (2005) 37 *New York University Journal of International Law and Politics* 995

Marsden, S, 'Public Participation in Transboundary Environmental Impact Assessment: Closing the Gap between International and Public Law' in B Jessup and K Rubenstein (eds), *Environmental Discourses in Public and International Law* (Cambridge University Press, 2012) 238

Mashaw, J L, 'Structuring a "Dense Complexity": Accountability and the Project of Administrative Law' (2005) 5(1) *Issues in Legal Scholarship* 1

Mason, M, and A Gupta, 'Transparency Revisited' in A Gupta and M Mason (eds), *Transparency in Global Environmental Governance: Critical Perspectives* (MIT Press, 2014) 321

Meredith, M, *Africa's Elephant: A Biography* (Hodder and Stoughton, 2001)

Michelson, K, 'Critical Approaches' in D Bodansky, J Brunnée and E Hey (eds), *The Oxford Handbook of International Environmental Law* (Oxford University Press, 2007) 262

Mitchell, R B, 'Compliance Theory: An Overview' in J Cameron, J Werksman and P Roderick (eds), *Improving Compliance with International Environmental Law* (Earthscan, 1996) 3

Mitchell, R B, 'Transparency for Governance: The Mechanisms and Effectiveness of Disclosure-based and Education-based Transparency Politics' (2011) 70 *Ecological Economics* 1882

Mofson, P, 'Protecting Wildlife from Trade: Japan's Involvement in the Convention on International Trade in Endangered Species' (1994) 3 *The Journal of Environment & Development* 91

Molina, M, et al, 'Reducing Abrupt Climate Change Risk using the Montreal Protocol and Other Regulatory Actions to Complement Cuts in CO_2 Emissions' (2009) 106(49) *Proceedings of the National Academy of Sciences* 20616

Montazeri, S, 'Protecting the Pachyderm: The Significance of Ivory Trade Regulation for African Elephant Conservation' (2013) 22 *Cardozo Journal of International & Comparative Law* 121

Montini, M, 'Procedural Guarantees in Non-Compliance Mechanisms' in T Treves et al (eds), *Non-Compliance Procedures and Mechanisms and the Effectiveness of International Environmental Agreements* (TMC Asser Press, 2009) 389

Nonet, P, and P Selznick, *Toward Responsive Law: Law & Society in Transition* (Transaction Publishers, 2001)

Oberthür, S, 'Options for a Compliance Mechanism in a 2015 Climate Agreement' (2014) 4 *Climate Law* 30

Oberthür, S, *Production and Consumption of Ozone-Depleting Substances 1986-1999: The Data Reporting System under the Montreal Protocol* (Deutsche Gesellschaft für Technische Zusammenarbeit, 2001)

Okafor, O C, 'Newness, Imperialism, and International Legal Reform in our Time: A TWAIL Perspective' (2005) 43 *Osgoode Hall Law Journal* 171

Ong, D M, 'The Convention on International Trade in Endangered Species (CITES, 1973): Implications of Recent Developments in International and EC Environmental Law' (1998) 10(2) *Journal of Environmental Law* 291

Ozone Secretariat, *Handbook for the Montreal Protocol on Substances that Deplete the Ozone Layer, Tenth Edition* (2016) <http://ozone.unep.org/sites/ozone/files/Publications/Handbooks/MP-Handbook-2016-English.pdf>

Ozone Secretariat, *Implementation Committee under the Non-Compliance Procedure of the Montreal Protocol on Substances that Deplete the Ozone Layer: Primer for Members* (2007) <http://ozone.unep.org/Publications/ImpCom_Primer_for_parties.pdf>

Payne, C, 'Introductory Note to Convention on International Trade in Endangered Species of Wild Fauna and Flora (CITES) Compliance Procedures' (2007) 46 *International Law Materials* 1174

Peel, J, 'New State Responsibility Rules and Compliance with Multilateral Environmental Obligations: Some Case Studies of How the New Rules Might Apply in the International Environmental Context' (2001) 20(1) *Review of European Community and International Environmental Law* 82

Peters, A, 'Compensatory Constitutionalism: The Function and Potential of Fundamental International Norms and Structures' (2006) 19 *Leiden Journal of International Law* 579

Pitea, C, 'NGOs in Non-Compliance Mechanisms under Multilateral Environmental Agreements: From Tolerance to Recognition' in T Treves et al (eds), *Civil Society, International Courts and Compliance Bodies* (TMC Asser Press, 2005) 204

Pitea, C, 'The Non-Compliance Procedure of the Aarhus Convention: Between Environmental and Human Rights Control Mechanisms' (2006) 16 *Italian Journal of International Law* 85

Rached, D H, 'Doomed Aspiration of Pure Instrumentality: Global Administrative Law and Accountability' (2014) 3(3) *Global Constitutionalism* 338

Rached, D H, *The International Law of Climate Change and Accountability* (PhD thesis, the University of Edinburgh, 2013)

Rajamani, L, 'Ambition and Differentiation in the 2015 Paris Agreement: Interpretative Possibilities and Underlying Politics' (2016) 65(2) *International and Comparative Law Quarterly* 493

Rajamani, L, 'Developing Countries and Compliance in the Climate Regime' in J Brunnée, M Doelle and L Rajamani (eds), *Promoting Compliance in an Evolving Climate Regime* (Cambridge University Press, 2012) 367

Rajamani, L, *Differential Treatment in International Environmental Law* (Oxford University Press, 2006)

Rajamani, L, 'The Changing Fortunes of Differential Treatment in the Evolution of International Environmental Law' (2012) 88(3) *International Affairs* 605

Rajamani, L, 'The Devilish Details: Key Legal Issues in the 2015 Climate Negotiations' (2015) 78(5) *Modern Law Review* 826

Rajamani, L, 'The Durban Platform for Enhanced Action and the Future of the Climate Regime' (2012) 61(2) *International and Comparative Law Quarterly* 501

Rajamani, L, J Brunnée and M Doelle, 'Introduction: The Role of Compliance in an Evolving Climate Regime' in J Brunnée, M Doelle and L Rajamani (eds), *Promoting Compliance in an Evolving Climate Regime* (Cambridge University Press, 2012) 1

Rajamani, L, and S Maljean-Dubois, 'The Present State of Research Carried Out by the English-speaking and French-speaking Sections – Report of the Directors of Studies' in S Maljean-Dubois and L Rajamani (eds), *Implementation of International Environmental Law* (Martinus Nijhoff Publishers, 2011) 107

Raustiala, K, 'NGOs in International Treaty-Making' in D B Hollis (ed), *The Oxford Guide to Treaties* (Oxford University Press, 2011) 150

Raustiala, K, *Reporting and Review Institutions in 10 Multilateral Environmental Agreements* (UNEP, 2001) <www.peacepalacelibrary.nl/ebooks/files/C08-0025-Raustiala-Reporting.pdf>

Raustiala, K, 'The "Participatory Revolution" in International Environmental Law' (1997) 21 *Harvard Environmental Law Review* 537

Reeve, R, *Policing International Trade in Endangered Species: The CITES Treaty and Compliance* (Earthscan, 2002)

Reeve, R, 'The Convention on International Trade in Endangered Species of Wild Fauna and Flora' in G Ulfstein, T Marauhn and A Zimmermann (eds), *Making Treaties Work: Human Rights, Environment and Arms Control* (Cambridge University Press, 2007) 134

Reeve, R, 'Wildlife Trade, Sanctions and Compliance: Lessons from the CITES Regime' (2006) 82(5) *International Affairs* 881

Romanin Jacur, F, 'Controlling and Assisting Compliance: Financial Aspects' in T Treves et al (eds), *Non-Compliance Procedures and Mechanisms and the Effectiveness of International Environmental Agreements* (TMC Asser Press, 2009) 419

Romanin Jacur, F, 'The Non-Compliance Procedure of the 1987 Montreal Protocol to the 1985 Vienna Convention on Substances that Deplete the Ozone Layer' in T Treves et al (eds), *Non-Compliance Procedures and Mechanisms and the Effectiveness of International Environmental Agreements* (TMC Asser Press, 2009) 11

Rose, G, 'Compliance with Multilateral Environmental Agreements' in L Paddock et al (eds), *Compliance and Enforcement in Environmental Law: Toward More Effective Implementation* (Edward Elgar, 2011) 3

Sabel, C F, and D G Victor, 'Governing Global Problems under Uncertainty: Making Bottom-up Climate Policy Work' (2015) *Climatic Change* 1

Sand, P H, 'Enforcing CITES: The Rise and Fall of Trade Sanctions' (2013) 22(3) *Review of European Community & International Environmental Law* 251

Sand, P H, 'International Protection of Endangered Species' (2017) 20 *Asia Pacific Journal of Environmental Law* (forthcoming)

Sand, P H, 'Sanctions in Case of Non-Compliance and State Responsibility: *pacta sunt servanda* – Or Else?' in U Beyerlin, P-T Stoll and W Rudiger (eds), *Ensuring Compliance with Multilateral Environmental Agreements: A Dialogue Between Practitioners and Academia* (Brill Martinus Nijhoff Publishers, 2006) 259

Sand, P H, 'Whither CITES? The Evolution of a Treaty Regime in the Borderland of Trade and Environment' (1997) 8 *European Journal of International Law* 29

Sandford, R, 'International Environmental Secretariats: A Case of Neglected Potential?' (1996) 16(1) *Environmental Impact Assessment Review* 3

Sands, P, 'Non-Compliance and Dispute Settlement' in U Beyerlin, P-T Stoll and W Rudiger (eds), *Ensuring Compliance with Multilateral Environmental Agreements: A Dialogue Between Practitioners and Academia* (Brill Martinus Nijhoff Publishers, 2006) 353

Schermers, H G, and N Blokker, *International Institutional Law: Unity within Diversity* (Brill, 5th ed, 2011)

Scott, C, 'Accountability in the Regulatory State' (2000) 27(1) *Journal of Law and Society* 38

Scott, K N, 'Non-Compliance Procedures and Dispute Resolution Mechanisms under International Environmental Agreements' in D French, M Saul and N D White (eds), *International Law and Dispute Settlement: New Problems and Techniques* (Hart Publishing, 2010) 225

Setear, J K, 'An Iterative Perspective on Treaties: A Synthesis of International Relations Theory and International Law' (1996) 139 *Harvard International Law Journal* 147

Shapiro, M, 'The Giving Reasons Requirement' (1992) *The University of Chicago Legal Forum* 179

Shapiro, M, *The Supreme Court and Administrative Agencies* (The Free Press, 1968)

Shigeta, Y, *International Judicial Control of Environmental Protection: Standard Setting, Compliance Control and the Development of International Environmental Law by the International Judiciary* (Wolters Kluwer, 2010)

Soltau, F, *Fairness in International Climate Change Law and Policy* (Cambridge University Press, 2009)

South, N, and T Wyatt, 'Comparing Illicit Trades in Wildlife and Drugs: An Exploratory Study' (2011) 32(6) *Deviant Behavior* 538

Stephens, T, *International Courts and Environmental Protection* (Cambridge University Press, 2009)

Stewart, R B, 'Accountability, Participation, and the Problem of Disregard in Global Regulatory Governance' (2008) <www.iilj.org/courses/documents/2008 Colloquium.Session4.Stewart.pdf>

Stewart, R B, 'Part I Courts, Institutions, and Access to Justice: Legitimacy and Accountability in Global Regulatory Governance: Global Administrative Law and Developing Countries' (2009) 1 *Jindal Global Law Review* 41

Stewart, R B, 'Remedying Disregard in Global Regulatory Governance: Accountability, Participation, and Responsiveness' (2014) 108 *American Journal of International Law* 211

Stewart, R B, M Oppenheimer and B Rudyk, 'Building Blocks for Global Climate Protection' (2013) 32 *Stanford Environmental Law Journal* 341

Sunstein, C R, 'Of Montreal and Kyoto: A Tale of Two Protocols' (2007) 31(1) *Harvard Environmental Law Review* 1

Tallberg, J, 'Orchestrating Enforcement: International Organizations Mobilizing Compliance Constituencies' in K W Abbott, P Genschel, D Snidal and B Zangl (eds), *International Organizations as Orchestrators* (Cambridge University Press, 2015) 166

Tanzi, A, and C Pitea, 'Non-Compliance Mechanisms: Lessons Learned and the Way Forward' in T Treves et al (eds), *Non-Compliance Procedures and Mechanisms and the Effectiveness of International Environmental Agreements* (TMC Asser Press, 2009) 569

Templeton, J, and P Kohler, 'Implementation and Compliance under the Minamata Convention on Mercury' (2014) 23(2) *Review of European Community and International Environmental Law* 211

Tomlinson, L, *Procedural Justice and the United Nations Framework Convention on Climate Change: Negotiating Fairness* (Springer, 2015)

Ulfstein, G, 'Depoliticizing Compliance' in J Brunnée, M Doelle and L Rajamani (eds), *Promoting Compliance in an Evolving Climate Regime* (Cambridge University Press, 2012) 418

Ulfstein, G, 'Dispute Resolution, Compliance Control and Enforcement in International Environmental Law' in G Ulfstein, T Marauhn and A Zimmermann (eds), *Making Treaties Work: Human Rights, Environment and Arms Control* (Cambridge University Press, 2007) 115

Ulfstein, G, 'Institutions and Competences' in J Klabbers, A Peters and G Ulfstein (eds), *The Constitutionalization of International Law* (Oxford University Press, 2009) 45

Ulfstein G, and J Werksman, 'The Kyoto Compliance System: Towards Hard Enforcement' in O S Stokke, J Hovi and G Ulfstein (eds), *Implementing the Climate Regime: International Compliance* (Earthscan, 2005) 39

Urueña Hernández, R F, 'Global Administrative Law and the Global South' in S Cassese (ed), *Research Handbook on Global Administrative Law* (Edward Elgar, 2016) 392

van Asselt, H, 'International Climate Change Law in a Bottom-up World' (2016) 26 *Questions of International Law* 5

Velazquez Gomar, J O, and L C Stringer, 'Moving Towards Sustainability? An Analysis of CITES' Conservation Policies' (2011) 21 *Environmental Policy and Governance* 240

Victor, D G, 'Enforcing International Law: Implications for an Effective Global Warming Regime' (1999) 10 *Duke Environmental Law and Policy Forum* 147

Victor, D G, *Global Warming Gridlock: Creating More Effective Strategies for Protecting the Planet* (Cambridge University Press, 2011)

Victor, D G, 'The Early Operation and Effectiveness of the Montreal Protocol's Non-Compliance Procedure', International Institute for Applied Systems Analysis (1996) <http://irps.ucsd.edu/dgvictor/publications/Victor_Article_1996_Early%20Operation%20and%20Effectiveness.pdf>

Victor, D G, 'The Operation and Effectiveness of the Montreal Protocol's Non-Compliance Procedure' in D G Victor, K Raustiala and E B Skolnikoff (eds), *The Implementation and Effectiveness of International Environmental Commitments* (MIT Press, 1998) 137

Voigt, C, 'The Compliance and Implementation Mechanism of the Paris Agreement' (2016) 25(2) *Review of European Community and International Environmental Law* 161

von Bogdandy, A, and M Goldmann, 'Taming and Framing Indicators: A Legal Reconstruction of the OECD's Programme for International Student Assessment' in K E Davis, A Fisher, B Kingsbury and S Engle Merry (eds), *Governance by Indicators: Global Power Through Quantification and Rankings* (Oxford University Press, 2012) 52

von Bogdandy, A, M Goldmann and I Vengke, 'From Public International to International Public Law: Translating World Public Opinion into International Public Authority' (2017) 28(1) *European Journal of International Law* 115

Waldron, J, 'Are Sovereigns Entitled to the Benefit of the International Rule of Law?' (2011) 22 *European Journal of International Law* 315

Weil, P, 'Towards Relative Normativity in International Law?' (1983) 88 *American Journal of International Law* 413

Werksman, J, 'Compliance and the Kyoto Protocol: Building a Backbone in a Flexible Regime' (1998) 9(1) *Yearbook of International Environmental Law* 48

Werksman, J, 'Compliance and Transition: Russia's Non-Compliance Tests the Ozone Regime' (1996) 56 *Zeitschrift für Ausländisches Recht und Völkerrecht* 750

Werksman, J, 'The Negotiation of a Kyoto Compliance System' in O Schram Stokke, J Hovi and G Ulfstein (eds), *Implementing the Climate Regime: International Compliance* (Earthscan, 2005) 17

Werner, W G, 'The Politics of Expertise: Applying Paradoxes of Scientific Expertise to International Law' in M Ambrus, K Arts, E Hey and H Raulus (eds), *The Role of 'Experts' in International and European Decision-Making Processes: Advisors, Decision Makers or Irrelevant Actors?* (Cambridge University Press, 2014) 44

Wettestad, J, 'Enhancing Climate Compliance – What are the Lessons to Learn from Environmental Regimes and the EU?' in O Schram Stokke, J Hovi and G Ulfstein (eds), *Implementing the Climate Regime: International Compliance* (Earthscan, 2005) 209

Wettestad, J, 'Monitoring and Verification' in D Bodansky, J Brunnée and E Hey (eds), *The Oxford Handbook of International Environmental Law* (Oxford University Press, 2007) 974

Wijnstekers, W, *The Evolution of CITES*, 9th edition (2011) <https://portals.iucn.org/library/efiles/documents/CITES-012-2011.pdf>

Yoshida, O, 'Procedural Aspects of the International Legal Regime for Climate Change: Early Operation of the Kyoto Protocol's Compliance System' (2011) 4 *Journal of East Asia and International Law* 41

Yoshida, O, 'Soft Enforcement of Treaties: The Montreal Protocol's Noncompliance Procedure and the Functions of Internal International Institutions' (1999) 10(1) *Colorado Journal of International Environmental Law and Policy* 95

Young, O R, 'Effectiveness of International Environmental Regimes: Existing Knowledge, Cutting-Edge Themes, and Research Strategies' (2011) 108(50) *Proceedings of the National Academy of Sciences* 19853

Zahar, A, 'A Bottom-Up Compliance Mechanism for the Paris Agreement', *Chinese Journal of Environmental Law* (forthcoming, 2017) <https://papers.ssrn.com/sol3/papers.cfm?abstract_id=2912560>

Zahar, A, *Climate Change Finance and International Law* (Routledge, 2017)

Zahar, A, 'Does Self-Interest Skew State Reporting of Greenhouse Gas Emissions? A Preliminary Analysis Based on the First Verified Emissions Estimates Under the Kyoto Protocol' (2010) 1 *Climate Law* 313

Zahar, A, *International Climate Change Law and State Compliance* (Routledge, 2015)

Zahar, A, 'Verifying Greenhouse Gas Emissions of Annex I Parties: Methods We Have and Methods We Want' (2010) 1 *Climate Law* 409

Zahar, A, J Peel and L Godden, *Australian Climate Law in Global Context* (Cambridge University Press, 2012)

B *Treaties*

Agreement for the Implementation of the Provisions of the United Nations Convention on the Law of the Sea of 10 December 1982 Relating to the Conservation and Management of Straddling Fish Stocks and Highly Migratory Fish Stocks, opened for signature 4 December 1995, 2167 UNTS 3 (entered into force 11 December 2001)

Basel Convention on the Control of Transboundary Movements of Hazardous Wastes and their Disposal 1989, opened for signature 22 March 1989, 1673 UNTS 57 (entered into force 5 May 1992)

Cartagena Protocol on Biosafety to the Convention on Biological Diversity, opened for signature 15 May 2000, 2226 UNTS 208 (entered into force 5 June 1992)

Convention for the Protection of World Cultural and Natural Heritage 1972, opened for signature 16 November 1972, 11 ILM 1358 (entered into force 17 December 1975)

Convention on Access to Information, Public Participation in Decision-Making and Access to Justice in Environmental Matters, opened for signature 25 June 1998, 38 ILM 517 (entered into force 30 October 2001)

Convention on Biological Diversity, opened for signature 5 June 1992, 1760 UNTS 79 (entered into force 29 December 1993)

Convention on International Trade in Endangered Species of Wild Fauna and Flora 1973, opened for signature 3 March 1973, 993 UNTS 243 (entered into force 1 July 1975)

Convention on Long Range Transboundary Air Pollution, opened for signature 13 November 1979, 18 ILM 1442 (entered into force 16 March 1983)

Convention on the Settlement of Investment Disputes between States and Nationals of Other States, opened for signature 18 March 1965, 575 UNTS 160 (entered into force 14 October 1966)

Convention on Wetlands of International Importance Especially as Waterfowl Habitat, opened for signature 2 February 1971, 11 ILM 963 (entered into force 21 December 1975)

Espoo Convention on Environmental Impact Assessment in a Transboundary Context, opened for signature 25 February 1991, 1989 UNTS 310 (entered into force 10 September 1997)

International Convention for the Regulation of Whaling 1946, opened for signature 2 December 1946, 161 UNTS 72 (entered into force 10 November 1948)

Kyoto Protocol to the United Nations Framework Convention on Climate Change, opened for signature 11 December 1997, 37 ILM 22 (entered into force 16 February 2005)

Minamata Convention on Mercury, opened for signature 10 October 2013, entry into force 16 August 2017

Montreal Protocol on Substances that Deplete the Ozone Layer 1987, opened for signature 16 September 1987, 1989 UNTS 28 (entered into force 1 January 1989)

Paris Agreement to the United Nations Framework Convention on Climate Change, opened for signature 12 December 2015 (entered into force 4 November 2016) (in UNFCCC, Report of the Conference of the Parties on its Twenty-First Session, Addendum, UN Doc FCCC/CP/2015/10/Add.1, 29 January 2016)

Protocol on Water and Health to the 1992 Convention on the Protection and Use of Transboundary Watercourses and International Lakes, opened for signature 17 June 1999, 2331 UNTS 202 (entered into force 4 August 2005)

United Nations Convention to Combat Desertification in Those Countries Experiencing Serious Drought and/or Desertification, Particularly Africa, opened for signature 17 June 1994, 1954 UNTS 3 (entered into force 14 October 1994)

United Nations Framework Convention on Climate Change, opened for signature 9 May 1992, 1771 UNTS 107 (entered into force 21 March 1994)

Vienna Convention for the Protection of the Ozone Layer, opened for signature 22 March 1985, 1513 UNTS 293 (entered into force 22 September 1988)

Vienna Convention on the Law of Treaties, opened for signature 23 May 1969, 1155 UNTS 331 (entered into force 27 January 1980)

C *Other primary sources*

1 CITES

CITES, CITES and Livelihoods, Resolution Conf 16.6 (Rev COP 17), Oct 2016

CITES, CITES Compliance Matters, prepared by the Secretariat for COP 17, Doc 23, Jul 2016

CITES, CITES Compliance Procedures, Resolution Conf 14.3, Jun 2007

CITES, CITES Strategic Vision: 2008–2020, Resolution Conf 16.3 (Rev COP 17), Oct 2016

CITES, CITES Trust Fund (CTL) Status of Contributions as of 30 June 2016 (in US Dollars), prepared by the Secretariat for COP 17, Doc 7.3, 8 Aug 2016

CITES, Compliance and Enforcement, Resolution Conf 11.3 (Rev COP 17), Oct 2016

CITES, Conservation of Cetaceans, Trade in Cetacean Specimens and the Relationship with the International Whaling Commission, Resolution Conf 11.4 (Rev COP 12), Nov 2002

CITES, Draft Decisions on Legal Acquisition Findings and Compliance Matters, COP 17, Com. II. 13, Oct 2016

CITES, Enforcement Matters, prepared by the Secretariat for COP 12, Doc 27, Nov 2002

CITES, Enforcement Matters, prepared by the Secretariat for COP 16, Doc 23, Mar 2013

CITES, Enforcement Matters, prepared by the Secretariat for the 45th Standing Committee meeting, SC45 Doc 11.2, Jun 2001

CITES, Enforcement Matters, prepared by the Secretariat for the 46th Standing Committee meeting, SC46 Doc 11.2, Mar 2002

CITES, Enforcement Matters, prepared by the Secretariat for the 47th Standing Committee meeting, SC47 Doc 10.2, Nov 2002

CITES, Establishment of Committees, Resolution Conf 11.1 (Rev COP 17), Oct 2016

CITES, Establishment of the Standing Committee of the Conference of the Parties, Resolution Conf 2.2, 1979 (Rev COP 15)

CITES, Financing and Budgeting of the Secretariat and of Meetings of the Conference of the Parties, Resolution Conf 7.2, Oct 1989 (Rev COP 12), Nov 2002

CITES, Guidelines for Compliance with the Convention, prepared by the Chairman of the Standing Committee Working Group on Compliance for COP 14, CoP14 Doc 23, 15 Jun 2000

CITES, Improving Transparency of Voting during Meetings of the Conference of the Parties, prepared by the Secretariat for the 62nd meeting of the Standing Committee, SC62 Doc 10.3, Jul 2012

CITES, Interpretation and Implementation of the Convention: Compliance and Enforcement, prepared by the Secretariat for COP 16, Doc 29 (Rev 1), Mar 2013

CITES, Interpretation and Implementation of the Convention – Review of Alleged Infractions, prepared for the Sixth Meeting of the Conference of the Parties Ottawa (Canada), Doc 6.19 (Rev), Jul 1987

CITES, Matters Relating to the Standing Committee: Report of the Chairman, prepared by the Secretariat for COP 11, Doc 11.8, Apr 2000

CITES, National Laws for the Implementation of the Convention, Resolution Conf 8.4 (Rev COP 15), Mar 2010

CITES, National Reports, Resolution Conf 11.17 (Rev COP 17), Oct 2016

CITES, Participation of Observers at Meetings of the Conference of the Parties, Resolution Conf 13.8 (Rev COP 16), Mar 2013

CITES, Proposal to Improve Transparency of Voting During Meetings of the Conference of the Parties, submitted by Denmark on behalf of the EU Member States for COP 16, Doc 4.2 (Rev 1), Mar 2013

CITES, Proposed Amendment to Rule 25 on Methods of Voting: Use of Secret Ballots, submitted by Chile and Mexico for COP 16, Doc 4.3 (Rev 1), Mar 2013

CITES, Recognition of the Benefits of Trade in Wildlife, Resolution Conf 8.3 (Rev COP 13), Oct 2004

CITES, Regional Representation on the Standing Committee, document prepared by Malawi on behalf of the African region for COP 9, Doc 9.7, Nov 1994

CITES, Review of Alleged Infractions, prepared by the Secretariat for COP 6, Doc 6.19 (Rev), Jul 1987

CITES, Review of Alleged Infractions and Other Problems of Enforcement of the Convention, prepared by the Secretariat for COP 8, Doc 8.19 (Rev), Mar 1992

CITES, Review of Alleged Infractions and Other Problems of Implementation of the Convention, prepared by the Secretariat for COP 9, Doc 9.22 (Rev), Nov 1994

CITES, Review of Alleged Infractions and Other Problems of Implementation of the Convention, prepared by the Secretariat for COP 10, Doc 10.28 (Rev), Jun 1997

CITES, Review of Alleged Infractions and Other Problems of Implementation of the Convention, prepared by the Secretariat for COP 11, Doc 11.20.1, Apr 2000

CITES, Review of Significant Trade in Specimens of Appendix II Species, Resolution Conf 12.8 (Rev COP 17), Oct 2016

CITES, Rules of Procedure of the Conference of the Parties, Rev COP 14, Jun 2007

CITES, Sponsored Delegates Project, document prepared by the European Union and Senegal for COP 17, Doc 8, Jun 2016

CITES, Staffing of the Secretariat, prepared by the Secretariat for COP 11, Doc 11.9.2, Apr 2000

CITES, Summary Report, prepared for the 45th Standing Committee meeting, SC45, Jun 2001

CITES, Summary Report, prepared for the 46th Standing Committee meeting, SC46, Mar 2002

CITES, Summary Report, prepared for the 47th Standing Committee meeting, SC47, Nov 2002

CITES, Trade in Elephant Specimens, Resolution Conf 10.10 (Rev COP 17), Oct 2016

2 The Kyoto Protocol, Paris Agreement and UNFCCC

Compliance Committee, Annual Report of the Compliance Committee to the Conference of the Parties Serving as the Meeting of the Parties to the Kyoto Protocol, Eighth Session, FCCC/KP/CMP/2012/6, 8 Nov 2012

Compliance Committee, Annual Report of the Compliance Committee to the Conference of the Parties Serving as the Meeting of the Parties to the Kyoto Protocol, Eleventh Session, FCCC/KP/CMP/2015/3, 18 Sept 2015

Compliance Committee, Annual Report of the Compliance Committee to the Conference of the Parties Serving as the Meeting of the Parties to the Kyoto Protocol, Ninth Session, FCCC/KP/CMP/2013/3, 1 Oct 2013

Compliance Committee, Annual Report of the Compliance Committee to the Conference of the Parties Serving as the Meeting of the Parties to the Kyoto Protocol, Seventh Session, FCCC/KP/CMP/2011/5, 3 Nov 2011
Compliance Committee, Annual Report of the Compliance Committee to the Conference of the Parties Serving as the Meeting of the Parties to the Kyoto Protocol, Sixth Session, FCCC/KP/CMP/2010/6, 8 Oct 2010
Compliance Committee, Annual Report of the Compliance Committee to the Conference of the Parties Serving as the Meeting of the Parties to the Kyoto Protocol, Third Session, FCCC/KP/CMP/2007/6, 26 Sept 2007
Compliance Committee, Decision on Preliminary Examination (Party Concerned: Croatia), CC-2009-1-2/Croatia/EB, 8 Sept 2009
Compliance Committee, Decision on Preliminary Examination (Party Concerned: Greece), CC-2007-1-2/Greece/EB, 22 Jan 2008
Compliance Committee, Decision on the Review and Assessment of the Plan Submitted Under Paragraph 2 of Section XV (Party Concerned: Greece), CC-2007-1-10/Greece/EB, 7 Oct 2008
Compliance Committee, Decision Under Paragraph 2 of Section X (Party Concerned: Bulgaria), CC-2010-1-17/Bulgaria/EB, 4 Feb 2011
Compliance Committee, Decision under Paragraph 2 of Section X (Party Concerned: Croatia), CC-2009-1-14/Croatia/EB, 8 Feb 2012
Compliance Committee, Decision under Paragraph 2 of Section X (Party Concerned: Greece), CC-2007-1-13/Greece/EB, 13 Nov 2008
Compliance Committee, Expert Advice: Greece, CC-2007-1-3/Greece/EB, 8 Feb 2008
Compliance Committee, Final Decision (Party Concerned: Greece), CC-2007-1-8/Greece/EB, 17 Apr 2008
Compliance Committee, Final Decision (Party Concerned: Ukraine), CC-2011-2-9/Ukraine/EB, 12 Oct 2011
Compliance Committee, Further Submission of Greece under Section X, Decision 27/CMP.1, CC-2007-1-7/Greece/EB, 9 Apr 2008
Compliance Committee, Further Written Submission from Ukraine, CC-2011-2-8/Ukraine/EB, 28 Sept 2011
Compliance Committee, Preliminary Finding (Party Concerned: Greece), CC-2007-1-6/Greece/EB, 6 Mar 2008
Compliance Committee, Report of the Review of the Annual Submission of Bulgaria Submitted in 2009, CC-2010-1-1/Bulgaria/EB, 17 Mar 2010
Compliance Committee, Report of the Review of the Initial Report of Greece, CC/ERT/IRR/2007/29, 31 Dec 2007
Compliance Committee, Written Submission from Croatia, CC-2009-1-5/Croatia/EB, 9 Oct 2009
Compliance Committee, Written Submission of Greece under Section X of Annex to Decision 27/CMP.1, CC-2007-1-5/Greece/EB, 26 Feb 2008
Decision 1/CP.16, Report of the Conference of the Parties on its Sixteenth Session, Held in Cancun from 29 November to 10 December 2010, FCCC/CP/2010/7/Add.1, 15 Mar 2011
Decision 2/CP.17, Outcome of the work of the Ad Hoc Working Group on Long-term Cooperative Action under the Convention, FCCC/CP/2011/9/Add.1, 15 Mar 2012
Decision 4/CMP.2, Compliance Committee, FCCC/KP/CMP/2006/10/Add.1, 4 Mar 2007

Decision 4/CMP.11, Revisions to the Guidelines for Review under Article 8 of the Kyoto Protocol, FCCC/KP/CMP/2015/8/Add.1, 29 Jan 2016, Annex I

Decision 14/CMP.7, Appeal by Croatia against a Final Decision of the Enforcement Branch of the Compliance Committee in Relation to the Implementation of Decision 7/CP.12, FCCC/KP/CMP/2011/10/Add.2, 15 Mar 2012

Decision 22/CMP.1, Guidelines for Review under Article 8 of the Kyoto Protocol, FCCC/KP/CMP/2005/8/Add.3, 30 Mar 2006

Decision 24/CP.7, Procedures and Mechanisms Relating to Compliance under the Kyoto Protocol, FCCC/CP/2001/13/Add.3, 21 Jan 2002

Decision 27/CMP.1, Procedures and Mechanisms Relating to Compliance under the Kyoto Protocol, FCCC/KP/CMP/2005/8/Add.3, 30 Mar 2006

UNFCCC Secretariat, Procedural Requirements and the Scope and Content of Applicable Law for the Consideration of Appeals under Decision 27/CMP.1 and Other Relevant Decisions of the Conference of the Parties Serving as the Meeting of the Parties to the Kyoto Protocol, as well as the Approach Taken by other Relevant International Bodies Relating to Denial of Due Process, Technical Paper, FCCC/TP/2011/6, 15 Sept 2011

UNFCCC Secretariat, Review of the Implementation of Commitments and of Other Provisions of the Convention: UNFCCC Guidelines on Reporting and Review, FCCC/CP/2002/8, 28 Mar 2003

3 The Montreal Protocol and the Ozone Convention

UNEP, *Assessment of the Basic Problems Confronting Countries with Economies in Transition in Complying with the Montreal Protocol: Report of the TEAP Ad-Hoc Working Group on CEIT Aspects* (UNEP, 1995)

UNEP, Decision III/2: Non-Compliance Procedure, UNEP/OzL.Pro.3/2, 21 Jun 1991

UNEP, Decision IV/5: Indicative List of Measures that might be taken by a Meeting of the Parties in Respect of Non-Compliance with the Protocol, Report of the Fourth Meeting of the Parties, UNEP/OzL.Pro.4/5, 25 Nov 1992

UNEP, Decision IV/5: Non-Compliance Procedure, UNEP/OzL.Pro.4/1, 25 Nov 1992

UNEP, Decision IX/31: Compliance with the Montreal Protocol by the Russian Federation, UNEP/OzL.Pro.9/12, 25 Sept 1997

UNEP, Decision VII/17: Compliance with the Montreal Protocol by Belarus, UNEP/OzL.Pro.7/12, 27 Dec 1995

UNEP, Decision VII/18: Compliance with the Montreal Protocol by Russia, UNEP/OzL.Pro.7/12, 27 Dec 1995

UNEP, Decision VII/19: Compliance with the Montreal Protocol by Ukraine, UNEP/OzL.Pro.7/12, 27 Dec 1995

UNEP, Decision VIII/25: Compliance with the Montreal Protocol by the Russian Federation, UNEP/OzL.Pro.8/12, 19 Dec 1996

UNEP, Decision X/10: Review of the Non-Compliance Procedure, UNEP/OzL.Pro.10/9, 3 Dec 1998

UNEP, Decision X/20: Compliance with the Montreal Protocol by Azerbaijan, UNEP/OzL.Pro.10/9, 3 Dec 1998

UNEP, Decision X/21: Compliance with the Montreal Protocol by Belarus, UNEP/OzL.Pro.10/9, 3 Dec 1998

UNEP, Decision X/22: Compliance with the Montreal Protocol by the Czech Republic, UNEP/OzL.Pro.10/9, 3 Dec 1998
UNEP, Decision X/23: Compliance with the Montreal Protocol by Estonia, UNEP/OzL.Pro.10/9, 3 Dec 1998
UNEP, Decision X/24: Compliance with the Montreal Protocol by Latvia, UNEP/OzL.Pro.10/9, 3 Dec 1998
UNEP, Decision X/25: Compliance with the Montreal Protocol by Lithuania, UNEP/OzL.Pro.10/9, 3 Dec 1998
UNEP, Decision X/26: Compliance with the Montreal Protocol by the Russian Federation, UNEP/OzL.Pro.10/9, 3 Dec 1998
UNEP, Decision X/27: Compliance with the Montreal Protocol by Ukraine, UNEP/OzL.Pro.10/9, 3 Dec 1998
UNEP, Decision X/28: Compliance with the Montreal Protocol by Uzbekistan, UNEP/OzL.Pro.10/9, 3 Dec 1998
UNEP, Decision XI/24: Compliance with the Montreal Protocol by Bulgaria, UNEP/OzL.Pro.11/10, 17 Dec 1999
UNEP, Decision XI/25: Compliance with the Montreal Protocol by Turkmenistan, UNEP/OzL.Pro.11/10, 17 Dec 1999
UNEP, Decision XIII/15: Data and Information Provided by the Parties to the 13th Meeting of the Parties in Accordance with Article 7 of the Montreal Protocol, UNEP/OzL.Pro.13/10, 26 Oct 2001
UNEP, Decision XIII/17: Compliance with the Montreal Protocol by the Russian Federation, UNEP/OzL.Pro.13/10, 26 Oct 2001
UNEP, Decision XIV/27: Request for Changes in Baseline Data, UNEP/OzL.Pro.14/9, 5 Dec 2002
UNEP, Decision XIX/24: Request for Change in Baseline Data by Turkmenistan, UNEP/OzL.Pro.19/7, 21 Sept 2007
UNEP, Decision XV/19: Methodology for Submission of Requests for Revision of Baseline Data, UNEP/OzL.Pro.15/9, 11 Nov 2003
UNEP, Decision XVI/31: Requests for Changes in Baseline Data, UNEP/OzL.Pro.16/17, 16 Nov 2004
UNEP, Decision XVIII/29: Request for Change in Baseline Data by Mexico, UNEP/OzL.Pro.18/10, 16 Nov 2006
UNEP, Decision XXIII/26: Non-compliance with the Montreal Protocol by the European Union, UNEP/OzL.Pro.23/11, 8 Dec 2011
UNEP, Decision XXIV/16: Requests for the Revision of Baseline Data by Algeria, Ecuador, Equatorial Guinea, Eritrea, Haiti, the Niger, the former Yugoslav Republic of Macedonia and Turkey, UNEP/OzL.Pro.24/10, 22 Nov 2012
UNEP, Decision XXV/11: Non-compliance with the Montreal Protocol by France, UNEP/OzL.Pro.25/9, 13 Nov 2013
UNEP, Decision XXVI/14: Requests for the Revision of Baseline Data by Libya and Mozambique, UNEP/OzL.Pro.26/10, 10 Dec 2014
UNEP, Decision XXVIII: Further Amendment to the Montreal Protocol, UNEP/OzL.Pro.28/CRP/10, 14 Oct 2016
UNEP, Decision XXVIII: Further Amendment to the Montreal Protocol, UNEP/OzL.Pro.28/CRP/11, 14 Oct 2016
UNEP, Decision XXVIII/10: Non-compliance by Israel with its Data and Information Reporting Obligations, UNEP/OzL.Pro.28/12, 15 Nov 2016
UNEP, Indicative List of Categories of Incremental Costs, Report of the Fourth Meeting of the Parties to the Montreal Protocol, UNEP/OzL.Pro.5/15, 25 Nov 1992

UNEP, Report of the Fourteenth Meeting of the Parties to the Montreal Protocol on Substances that Deplete the Ozone Layer, UNEP/OzL.Pro.14/9, 5 Dec 2002

UNEP, Report of the Fourth Meeting of the Implementation Committee, UNEP/OzL.Pro/ImpCom/4/2, 6 Oct 1992

UNEP, Report of the Fourth Meeting of the Parties, UNEP/OzL.Pro.4/15, 25 Nov 1992

UNEP, Report of the Implementation Committee under the Non-Compliance Procedure for the Montreal Protocol on the Work of its Eleventh Meeting, UNEP/OzL.Pro/ImpCom/11/1, 14 Sept 1995

UNEP, Report of the Implementation Committee under the Non-Compliance Procedure for the Montreal Protocol on the Work of its Twentieth Meeting, UNEP/OzL.Pro/ImpCom/20/4, 9 Jul 1998

UNEP, Report of the President of the Implementation Committee on the Report of the Secretariat on Information Provided by the Parties in Accordance with Articles 4, 7 and 9 of the Montreal Protocol and on the Functioning of the Implementation Committee, UNEP/OzL.Pro.7/12, 27 Dec 1995

UNEP, Report of the Seventh Meeting of the Parties to the Montreal Protocol on Substances that Deplete the Ozone Layer, UNEP/OzL.Pro.7/12, 27 Dec 1995

UNEP, Report of the Third Meeting of the Ad-Hoc Working Group of Legal and Technical Experts on Non-Compliance with the Montreal Protocol, UNEP/OzL.Pro/WG.3/3/3, 9 Nov 1991

UNEP, Report of the Thirty-Ninth Meeting of the Implementation Committee, UNEP/OzL.Pro/ImpCom/39/7, 11 Nov 2007

UNEP, Report of the Twelfth Meeting of the Parties to the Montreal Protocol, UNEP/OzL.Pro.12/9, 10 Jan 2001

UNEP, Report on the Work of the Ad Hoc Working Group of Legal and Technical Experts on Non-compliance with the Montreal Protocol, UNEP/OzL.Pro/WG.4/1/3, 18 Nov 1998

UNEP, 'Rules of Procedure for Meetings of the Conference of the Parties to the Vienna Convention and Meetings of the Parties to the Montreal Protocol', *Handbook for the Vienna Convention for the Protection of the Ozone Layer* (7th ed, 2006) <http://ozone.unep.org/Publications/VC_Handbook/Section_3_Rules_of_Procedure/index.shtml>

D *Other*

CITES, 'CITES World: Official Newsletter of the Parties' (June 1999) Issue 3, 2 <www.cites.org/eng/news/world/3.pdf>

CITES, 'Conference of the Parties' (2017) <https://cites.org/eng/cop/index.php>

CITES, 'Countries Currently Subject to a Recommendation to Suspend Trade' (last updated 1 Feb 2017) <http://cites.org/eng/resources/ref/suspend.php>

CITES, 'Decisions of the Conference of the Parties to CITES in Effect after the 17th Meeting' (2017) <https://cites.org/eng/dec/index.php>

CITES, 'Japan' (2017) <www.cites.org/eng/cms/index.php/component/cp/country/JP>

CITES, 'Participation of Non-State Actors in CITES' (Presentation at UNECE/Forest Europe Seminar on Developing an Effective International Legal Instrument, Geneva, Switzerland, 7 September 2010) <www.unece.org/fileadmin/DAM/timber/meetings/20100907/6_Marceil_Yeater.pdf>

CITES, 'The CITES Species' (2017) <www.cites.org/eng/disc/species.php>

CITES, 'The Standing Committee' (2017) <www.cites.org/eng/com/sc/index.php>

CITES, 'What is CITES?' (2017) <www.cites.org/eng/disc/what.php>

Global Ozone Observing System, 'Global Observing System Information Centre' (2017) <www.ncdc.noaa.gov/gosic>

IISD Reporting Services, 'Summary of the Fourteenth Meeting of the Conference of the Parties to the Convention on International Trade in Endangered Species of Wild Fauna and Flora' (18 June 2007) 21(61) *Earth Negotiations Bulletin* 2 <http://enb.iisd.org/download/pdf/enb2161e.pdf>

IISD Reporting Services, 'Summary of the Seventeenth Meeting of the Conference of the Parties to the Convention on International Trade in Endangered Species of Wild Fauna and Flora' (8 October 2016) 21(97) *Earth Negotiations Bulletin* <www.iisd.ca/download/pdf/enb2197e.pdf>

IISD Reporting Services, 'Summary of the Sixteenth Meeting of the Conference of the Parties to the Convention on International Trade in Endangered Species of Wild Fauna and Flora' (18 March 2013) 21(83) *Earth Negotiations Bulletin* <http://www.iisd.ca/vol21/enb2183e.html>

International Tribunal of the Law of the Sea, *Responsibilities and Obligations of States Sponsoring Persons and Entities with respect to Activities in the Area (Advisory Opinion),* Seabed Disputes Chamber, Case No 17, 1 Feb 2011

IUCN, 'About' (2017) <www.iucn.org/about>

IUCN/TRAFFIC, 'IUCN/TRAFFIC Analyses of the Proposals to Amend the Appendices for CITES CoP16' (2013) <https://cmsdata.iucn.org/downloads/iucn_traffic_analyses_of_proposals_cites_cop16_2012_1.pdf>

Ministry of Foreign Affairs of Japan, 'The Convention on International Trade in Endangered Species of Fauna and Flora (CITES)' (2016) <www.mofa.go.jp/policy/environment/convention/cites.html>

Multilateral Fund for the Implementation of the Montreal Protocol, 'Welcome to the Multilateral Fund for the Implementation of the Montreal Protocol' (2017) <www.multilateralfund.org/default.aspx>

Rio+20 Conference, 'Our Common Vision' (2012) <www.un.org/disabilities/documents/rio20_outcome_document_complete.pdf>

TRAFFIC, 'About Traffic' (2017) <www.traffic.org/overview/>

TRAFFIC, 'Traffic and CITES' (2017) <www.traffic.org/cites/>

UNEP, 'Agenda 21' (1992) <https://sustainabledevelopment.un.org/content/documents/Agenda21.pdf>

UNEP, 'The Kigali Amendment to the Montreal Protocol: Another Global Commitment to Stop Climate Change' (2016) <http://web.unep.org/kigali-amendment-montreal-protocol-another-global-commitment-stop-climate-change>

UNEP, 'Treaties and Decisions' (2017) <http://ozone.unep.org/en/treaties-and-decisions>

UNEP Ozone Secretariat, 'Adjustments to the Montreal Protocol' (2017) <http://ozone.unep.org/en/handbook-montreal-protocol-substances-deplete-ozone-layer/27594>

UNEP Ozone Secretariat, 'Key Achievements of the Montreal Protocol to Date' (2012) <http://ozone.unep.org/new_site/en/Information/Information_Kit/Key_achievements_of_the_Montreal_Protocol_2012.pdf>

UNEP Ozone Secretariat, 'The Vienna Convention for the Protection of the Ozone Layer' (2017) <http://ozone.unep.org/en/treaties-and-decisions>

UNFCCC, 'About UNFCCC' (2017) <http://newsroom.unfccc.int/about/>

UNFCCC, 'Compliance under the Kyoto Protocol' (2017) <http://unfccc.int/kyoto_protocol/compliance/items/2875.php>.

UNFCCC, 'Initial Reports under Article 7, paragraph 4, of the Kyoto Protocol and Initial Review Reports' (2017) <http://unfccc.int/national_reports/initial_reports_under_the_kyoto_protocol/first_commitment_period_2008-2012/items/3765.php>

UNFCCC, 'UNFCCC Roster of Experts' (2017) <www4.unfccc.int/sites/roe/Pages/Home.aspx>

UNFCCC, 'Webcasts and Videos: Compliance Committee' (2017) <https://unfccc.int/press/multimedia/webcasts/items/5862.php>

UNFCCC Webcast, 'The Third Meeting of the Enforcement Branch: Bonn, 4 March – 6 March 2008' (2008) <http://unfccc.int/resource/webcast/compliance/enf3/templ/ovw_unfccc7729.html>

United Nations General Assembly, Rio Declaration on Environment and Development (Annex I), A/CONF.151/26 (vol 1), 14 Jun 1992, reprinted in 31 ILM 874

United Nations Treaty Collection, 'Chapter XXVII Environment: 7.c Doha Amendment to the Kyoto Protocol' (2017) <https://treaties.un.org/pages/ViewDetails.aspx?src=TREATY&mtdsg_no=XXVII-7-c&chapter=27&clang=_en>

World Commission on Environment and Development, 'Report of the World Commission on Environment and Development: Our Common Future' (1987) <www.un-documents.net/wced-ocf.htm>

World Meteorological Organization and United Nations Environment Programme, *Assessment for Decision-Makers: Scientific Assessment of Ozone Depletion: 2014*, World Meteorological Organization Global Ozone Research and Monitoring Project, Report No 56, Geneva, Switzerland <http://ozone.unep.org/Assessment_Panels/SAP/SAP2014_Assessment_for_Decision-Makers.pdf>

Index

For Product Safety Concerns and Information please contact our EU representative GPSR@taylorandfrancis.com Taylor & Francis Verlag GmbH, Kaufingerstraße 24, 80331 München, Germany

Batch number: 10725973

Printed by Printforce, the Netherlands